HYDROLOGY AND WATER RESOURCES IN TROPICAL AFRICA

DEVELOPMENTS IN WATER SCIENCE, 8

advisory editor

VEN TE CHOW

Professor of Hydraulic Engineering
Hydrosystems Laboratory
University of Illinois
Urbana, Ill., U.S.A.

FURTHER TITLES IN THIS SERIES

HYDROLOGY
AND WATER RESOURCES
IN TROPICAL AFRICA

by
JAROSLAV BALEK
Senior Researcher, Institute of Hydrodynamics, Academy of Science, Prague

ELSEVIER SCIENTIFIC PUBLISHING COMPANY

AMSTERDAM — OXFORD — NEW YORK — 1977

Published in co-edition with
SNTL Publishers of Technical Literature, Prague

Distribution of this book is being handled by the following publishers

for the U.S.A. and Canada
Elsevier North-Holland, Inc.
52 Vanderbilt Avenue
New York, N.Y. 10017

for the East European Countries, China, Northern Korea, Cuba, Vietnam and Mongolia

SNTL Publishers of Technical Literature, Prague

for all remaining areas
Elsevier Scientific Publishing Company
335 Jan van Galenstraat
P.O. Box 211, Amsterdam, The Netherlands

Library of Congress Cataloging in Publication Data

Balek, J
   Hydrology and water resources in tropical Africa.

   (Developments in water science ; 8)
   Includes bibliographies.
   1. Hydrology--Africa--Sub-Saharan. 2. Water-
supply--Africa, Sub-Saharan. I. Title. II. Series.
GB800.B25   553.7'0967   77-5566
ISBN 0-444-99814-4

Printed in Czechoslovakia

FOREWORD

by J. M. K. Dake, B. Sc. (Eng), M. Sc. Tech, Sc. D,
Professor of Civil Engineering,
University of Zambia, Lusaka

Doctor Jaroslav Balek could hardly have given himself a more difficult task than the one tackled in his book on Hydrology and Water Resources in Tropical Africa. Writing a Foreword to the book is almost equally demanding.

Tropical Africa, as is evident in the various chapters of the book, is a very complex geographical and climatic region. Categorization of its relief, climate, vegetation, geology, etc, and their interaction with the ecology, hydrology and morphology of its river basins, lakes, dambos, swamps, etc, is a near impossible assignment, even if sufficient data were available; non-availability of sufficiently reliable and scientifically oriented data makes the job even much more difficult.

Consequently, Dr. Balek's effort should not be judged by preciseness or lack of it of the information provided, but rather by its boldness in collecting and collating various scientific information and speculations and relating them to his personal experiences and those of others to give a general picture.

Dr. Balek, obviously, has not sought to write a handbook on Tropical African Hydrology and Water Resources. Neither the geographer nor the hydrological engineer will find much detailed information in the book for planning and design purposes.

However, the research geographer, hydrologist, water resources engineer and the environmentalist will find the book very useful. Dr. Balek has drawn attention to the many hydrological, climatological and ecological factors which should form the basis for a scientific study of water resources in the tropics. He has provided, at many places, guides and hints for a correct approach to such studies. Young African researchers and academics who generally have to tackle such problems without much superior guiding experience will find the various hints an invaluable asset.

PREFACE

This book deals with the special field of regional hydrology. It is concerned with the hydrological cycle as existing between the Tropics of Cancer and Capricorn on the African continent. Regarding the close relationship between hydrology and ecology on one side and hydrology and engineering science on the other, the intention was to present points of view common to engineering hydrology, ecology and geography.

A great economic potential of the tropics, particularly in the developing countries, is bound up with existing natural resources, water being one of the most significant. A frequent experience is that the research results and theoretical conclusions obtained in various fields of hydrology in the moderate regions, cannot be applied directly to the tropics. In many cases the theory needs to be extended or even rebuilt. Therefore, those problems of the hydrological cycle which should be taken into account as being special for the tropics, have been emphasized. On the contrary, common problems of general hydrology, however relevant to the topic, have been minimized, when data are available from other sources.

Basically, the book is based on the results of my own experience, but many conclusions presented have been reached by other authors. A list of the references is attached to the end of each chapter.

Geographical terminology in Africa is in a process of rapid development in the same way as the whole continent. British, French or local transcription commonly used until 1975 have been used in this book. This book could never have been written without the direct or indirect assistance of many people. I particularly appreciate the cooperation of my colleagues, both engineers and technicians from the Water Resources Unit, National Council for Scientific Research of Zambia, with whom I spent several years working in the field of tropical hydrology. I am also indebted to Mrs. J. Kulveitová, Institute of Hydrodynamics, Prague, Czechoslovakia, for the illustrations and graphic work in this book.

Last but not least, I wish to express my appreciation of the continuous effort of people who indirectly contributed to this book when exploring, surveying and gauging in the exciting world of African tropical rivers, lakes and swamps, from the beginning of this century.

J. Balek

CONTENTS

8

1. HISTORY OF AFRICAN HYDROLOGY
AND WATER RESOURCES PRIOR TO 1900 A. D.

L. S. B. Leakey:
"... I should like to stress the fact that if man lived as he did,
close to water, he always had food. The dependence on water
served man well, but it restricted him terribly and the time
came when two things happened in praehistoric times: first man
learned to make and control fire and secondly he learned to
make vessels of skin, ostrich egg shell and so on, to carry and
to store water. This meant immediately that he could extend
his habitat to an incredible degree, he could move outwards
away from permanent waters, carrying water with him. It
means he could live in more comfortable places like caves and
rock shelters. In Africa, however, even when man started to
conserve water in skins and ostrich shells he still could never
move very far from water and water today still dominates the
tropical African scene."

1.1 AFRICAN MAN, LAKES AND RIVERS

Since ancient times African man has been dependent on the natural resources
of the continent and the fluctuation of the hydrological cycle has influenced his life
and development to a great extent. It is very likely Africa where the earliest stages
of man's cultural and biological development originated. Several hundred man-ape
fossils believed to be several million years old have been found in East and South
Africa. In 1974 fossils of an ancestor of man were found in the Awash valley,
Ethiopia, which are believed to be 3 − 4 millions years old. A common predecessor
of apes and man, Proconsul Africanus, lived in Africa twenty million years ago
and Kenyapithecus, a creature without tools and perhaps the last ancestor of man,
was found in East Africa. A famous locality, Olduvai, in Tanzania produced
Zinjanthropus, and Homo Habilis, a creature using the first tools made from bones
and pebbles, also originated in East Africa only one million years ago. It can be
seen from Fig. 1.1 that most of the prehistoric localities in tropical Africa are spread
out along the great rivers and lakes (see enclosure).

The earliest stages of man's existence in Africa were influenced by the fluctuation
of wet and dry periods, the so-called interpluvials and pluvials. Wayland [29]
identified the main periods of increased rainfall, separated by periods of drier
climate, at Lake Victoria. Nilsson [20] together with Leakey, found two wet post-
pleistocene phases called Makalian and Makuran; two wet periods in the Late
Pleistocene, called Gamblian and Nakuran, were also traced and a Kamasian period

in the Middle Pleistocene and Kageran in the Early Pleistocene were also identified. Accordingly, rich cultural history was correlated by Leakey with the periods of pluvials and faunal development. The reconstruction of past environment has therefore been an essential part of Quarternary archaeology in Africa where the ecological zones can be well identified.

The regions of equal summer and winter rainfall probably remained the same throughout the Quaternary and only small changes occurred. Instead of changes in the ecological systems, there was some shifting of the vegetational and faunal zones and ecological boundaries. However, these shifts were conceivably extensive. For instance, during Late Acheulian and Aterian times, the Sahara desert was invaded by Mediterranean flora down to the southern limits. There must have been more temperate conditions and an annual rainfall of at least 25 inches in the area that is today true desert. Clark and Summers [6, 25] in Southern Rhodesia and Cooke [5] in South Africa reconstructed hypothetical rainfalls and vegetational patterns at the maxima of pluvial and interpluvial conditions.

Variable hydrological conditions also dominantly affected the distribution of human population. The drier conditions, still existing at the close of the Pleistocene, produced wide population movements and supported an interchange of improved agricultural techniques. On the other hand, more favourable conditions, such as existed during the Gamblian pluvial period, might have had a stabilising effect on the existing settlements. At some time the whole of Africa was populated by hunters, as can be traced from the hunting scenes on the walls of caves in the Sahara and other parts of Africa. Some regions became drier and man and animals sought other sources of water. At that time man was very likely learning two other ways of getting food, namely the herding of domesticated animals and cultivation of the soil. The hunters survived in the lush grasslands and the tropical rainforest. The herders remained nomadic, seeking pastures, while the cultivators were tied, as in the Nile valley, by water sources. The presence of a permanent water supply made possible the production of surplus food and resulted in trade and paid labour. The river became also the means of trade communication and a favourable place for the creation of centralized rule. An example of this was given by Turnbull [26] in the Sudan where the cultivators, Shilluks, built their homes on high land connected by ridges so that during floods, the villages were not completely isolated. Settlements were divided into hamlets and each settlement had its own headman elected by the leaders of the hamlets and confirmed in the appointment by "the divine king". Farther up the Nile, other tribes, such as the Dinka and Nuer had no centralization because both were more pastoral with more isolated and remote settlements.

The Kalahari Bushman can serve as another example of adaptation to the existing water supply. To an outsider it seems almost impossible to live in the Kalahari area of arid scrubland with occassional thorn bushes. However the Bushman's knowledge of hydrogeology is extraordinary. In areas where one would expect no water within a hundred miles, he will walk a few feet, dig down into the sand and suck the water out of it. Sometime the Bushmen fill ostrich egg shells with the excess water and hide them below the sand for less favourable times.

1.2 NILE BASIN

The oldest known hydrological records are from the African continent and the history of its management of water resources is among the oldest in the world. As stated by Biswas [3], one of the most ancient records is a drawing of an imperial macehead held by the protodynastic King Scorpion when participating in the ceremony of cutting an irrigation ditch. His reign is dated at about 3200 B. C.

The section of the river Nile for which historical hydrological records exist is outside the tropical boundaries, but the river itself is one the most remarkable products of a tropical environment. Herodotus, who came to Egypt in the fifth century B.C., found still in existence there the oldest Nile dam, built south of Memphis, under the rule of Pharaoah Menes about 3000 B. C. The oldest dam in the world at present is in the Wadi el Garawi, only 18 miles from Cairo, probably built between 2950 – 2750 B. C. In the Middle Kingdom (2160 – 1788 B. C.) several artificial lakes were constructed, presumably by order of Amenemhet III. The simple water wheels have remained unchanged since ancient times.

Herodotus made a very appropriate comparison, in calling Egypt "Gift of the Nile". The relatively narrow strip of inundated, and therefore cultivated soil on both banks of the Nile has always been bounded by the desert. Since the end of prehistoric times an increase in agriculture can be noted in the so-called inundated zone and it made a surprising leap at the beginning of the Old Kingdom, about 2800 B. C.

Life in the whole country has always been dependent on the seemingly uncontrollable whims of this the greatest river of Africa. A sufficient quantity of water has meant rich crops while water shortage has resulted in poor crops and famine. The effort to unify the disorganised system of water regulation and to build irrigation tracts was a very important, perhaps decisive factor in the emergence and growth of state power. A number of examples can be quoted demonstrating the immense importance ascribed to the Nile by the ancient Egyptians. The Nile, personified by the god Hapy, was a significant figure in their religious imagery. The Nile floods, in particular, furnished a reason for complicated religious rites and theological doctrines as well as for very intensive observation of all astronomic phenomena and the corresponding occurrences on the earth which were related to the floods.

The pragmatic Egyptians promptly made the best possible use of the results of their findings. It was prolonged and thorough observation of the Nile that enabled them to predict rich or poor crops and, on the basis of the predictions, the ruler could regulate taxes etc. Verner [28] concludes that the computation of the hypothetical fluctuation of the Nile's water volume together with the present knowledge of Egyptian history, may result in a possible correlation between periods of increased water volume and periods of political, economical and cultural development of the whole region.

In Genesis, Chap. 41, v. 26−37 tells the biblical story of Joseph interpreting Pharaoah's dream; there is an indication that the number seven does not occur by accident but that it has a very specific meaning in the fluctuation of the river. Another such sign is what is probably the very first hydrologic record engraved on a stele on the island Sehel, stating that "The Nile has not come for seven years" [2]. A number of other records which give information on the height of Nile floods from Semna and Kumma near the Second cataract for example, have also been preserved [23]; other records, from Karnak, are described by Legrain [16]. Such data however, are not comprehensive enough to help us ascertain the periodicity of the river of other phenomena.

The height of Nile floods was measured by means of nilometres, the most famous situated on the island of Rauda [4]. The origin of the nilometre was once a subject for polemics among Egyptologists as to whether it originated in ancient Egypt or was constructed by the Arabs after the conquest of Egypt. The German Egyptologist Sethe [24] considered the nilometre to be a sanctuary of the god Hapy, existing already in Pharaonic Egypt. Later the British Egyptologist Gardiner [14] arrived at a different conclusion, but in 1953 the French Egyptologist Drioton [9] confirmed Sethe's hypothesis. Actually, there must have been many more nilometers but only a few have been historically documented. Omar Toussoun lists thirty-one in his study. A number of them were built by the Arabs who introduced better order into the annual records on the stages of the Nile. According to Popper [22] there must have been a systematically written record of the Nile because at the time of floods the culmination of the river level was announced together with that of the preceding year. Unfortunately, no original record has been found and we depend on the manuscripts of the Arab historians Taghri Birdi and Al-Hijazi who have recorded maxima an minima sequences over 849 years commencing with 622 A. D. [1].

The history of the search for the sources of the river is remarkable. According to Ptolemy, an Egyptian, Diogenes (second century A. D.) reported two large interior lakes in tropical Africa and a range of snow-capped mountains in the region where the Nile rose. Diogenes called them the Mountains of the Moon. Ptolemy writes that the snow melted and fed two lakes and from each lake there flowed two rivers, later joined by the Nile. It is still not certain whether Diogenes visited Mt. Kenya and Kilimanjaro or Ruwenzori, which resemble the visible surface of the moon. Both mountains, the Mountains of the Moon and the two lakes, are plotted on Ptolemy's map of 150 A. D. as sources of the river. The Blue Nile is also shown on the map.

Very little can be said about the Nile after the second century A. D. and up to the nineteenth century, when the educational and economic climate in Europe encouraged travellers to seek an answer to the geographical problems of Africa. Their attention was drawn most strongly to the White Nile which was believed to flow from the main sources of the river. In 1848 Ludwig Krapf and J. Rebmann, missionaries at Mombasa, gathered information on the African interior which corresponded with Ptolemy's description. They reported the existence of two snow-capped mountains and they saw Kilimanjaro. They also reported the Sea of Ujiji, which was discredited by European geographers, who were still discussing in 1856 whether the Nile originated from the fountains of Herodotus.

In 1855, Captain R. F. Burton proposed an expedition to the Sea of Ujiji and was entrusted by the Royal Geographical Society with looking for the Mountains of the Moon and sources of the Nile. In the company of J. H. Speke, he learned that the Sea of Ujiji is in reality three huge lakes; the Nyasa, Tanganyika and Victoria. In 1858 they proceeded to Lake Tanganyika and Burton became the first European to see the lake as Speke was suffering from opthalmia. Later, Burton caught malaria, and Speke alone reached Lake Victora the same year, although his ailing eyes prevented him from seeing much.

Contrary to Speke, Burton was not convinced that Lake Victoria is the source of the Nile. Speke was requested by the Royal Geographical Society to prove the assumption. Accompanied by J. A. Grant, Speke returned to the Lake and in 1862 he saw the Nile Falls at the western side. On the way north they met Samuel White Baker and his wife, who were also following the stream in search of the Nile's source. Lake Albert was reached in 1864 and in 1866 Livingstone, still unconvinced that the river's true sources had been found, speculated that Lake Nyasa might drain into Lake Tanganyika, which could be linked with Lake Albert and with the Nile. He believed that the sources of the Congo and the Nile could be identical and in 1872, one year after he was found at Ujiji by H. M. Stanley, he died at Bangweulu swamps when still searching for the fountains.

The actual source of the river was not made known until 1937, when a German explorer, B. Waldecker, traced the southernmost tributary of the Kagera river, flowing into Lake Victoria.

The Blue Nile was explored with great difficulties. It was believed that it flowed from sacred springs and only the priests from the neighbouring church could take water from it. The sources were first visited by a Portuguese Jesuit, Pedro Paez in the seventeenth century. In 1770, the headwaters were reached by James Bruce who became so friendly with the ruler that he was appointed Governor there.

For a long distance from Tana, the Blue Nile flows through a canyon, in some places 1200 meters deep. As late as in 1905, McMillan's expedition was able to follow only about two fifths of the river. McMillan tried to run the river in a steel boat but was almost drowned at the very outset. In 1923, the top of the canyon was travelled by an expedition led by R. E. Cheesman.

Another success in the hydrographical surveying of this part of Africa was completed in 1888 when Count Teleki and lieutenant Von Höhnel discovered Lake Rudolph.

1.3 THE RIVER NIGER AS A TRADING CENTRE

The Middle Niger has been an important centre for the West African market since the Christian era. The trade centres established in the basin connected the West African coast with the Mediterranean and the Sudan. The first trade centres have been traced by historians back to the first centuries of the Christian era when the Songhai kingdom already existed between Gao and Bussa. The kingdom had not reached its zenith until the fourteenth century when it became the real centre for trade between Tunis and Egypt on one side and the West coast on the other.

From the hydrographical point of view, the Niger has been a subject for speculation from the times of Ptolemy. Once it was considered to be identical with the Congo, and at another time as a separate river which disappeared somewhere in the middle of the desert. Leo Africanus suggested that the river flowed from an interior source into the ocean, but in a westwards direction.

In 1788, the Association for Promoting the Discovery of the Interior of Africa was established and one of its tasks was the search for the headwaters of the Niger and its hydrography. John Ledyard, an American, was asked by the Association to travel across Africa near to the Niger river, however, he died deep inland before he was able to complete his mission. Then Major Daniel Francis Houghton was sent to the basin of the Ike Gambia river, once believed to be connected with the Niger. He too, died when on expedition, but at least he proved that the original idea of Leo Africanus was not correct. The sources of the Gambia and Senegal rivers were discovered by a Frenchman, Gaspard Mollien, in 1818.

A Scotish doctor, Mungo Park, was requested by the Association "... to ascertain the course and, if possible, the river and its termination". He reached the Niger at Ségou in 1796, very much impressed: "Looking forwards, I saw with infinite pleasure the great object of my mission; the long sought for, majestic Niger, glittering in the morning sun, as broad as the Thames at Westminster, and flowing slowly eastward."

Mungo Park was able to follow the river for only six days, concluding from reports he obtained from the Negro merchants at Timbuctoo and Haussa, that it flows to "the world's end".

Another attempt was made by F. C. Hornemann who believed that the Ike Niger and Nile were identical rivers and started his search by following the caravan trail from Cairo. He died before reaching the river.

Early in the nineteenth century, a theory was put forward by British geographers that the Niger flows into the swamp of Timbuktoo and evaporates there. Christian Gottlieb Reichard, had the idea that the Niger flowed into the Bight of Benin, but his theory was discredited during his time. In 1805, the British Government sponsored an expedition of 45 explorers, led by Mungo Park. One by one, they perished and then the rest of the expedition, including Mungo Park, was lost somewhere at Bussa.

It is not known for how long, but definitely at the beginning of nineteenth century, the Muslim rulers of Haussaland knew the route of the Niger to the ocean. Muhammed Bello at Sokoto sketched a map of the Niger for Lieutenant Hugh Clapperton already in 1824. He explained how easily the Middle Niger trade centers could be reached by European ships. However, he refused to permit Clapperton to prove it. Clapperton travelled first to Lake Chad, accompanied by Dr. Walter Oudney and Major Dixon Denham, and then visited the Bight of Benin. From there, he visited Muhammed's Bello Empire travelling overland, after attempting to prove that the Oil Rivers at Brass are identical with the Niger. He died at Sokoto in 1827 and Richard Lemon Lander, his former servant, tried to follow the Niger downstream. A local ruler refused to allow him to complete the trip; in 1830 he returned with his brother John and after reaching Bussa, they sailed down the river in two small canoes. They reached Brass and proved that the Oil Rivers and the Niger are one stream.

In 1832–1833 two steamships, under the command of Lander, explored the lower part of the Niger up to the confluence with the river Benue and then followed the Benue itself. Forty of the fortyfour members of the expedition including the commander, died during the trip. In 1841 another big expedition steamed higher up the Niger than Lander; again a large number of the explorers died of fever. Finally, Dr. William Balfour Baikie guided another ship upstream of the Niger and Benue and opened the increasingly significant trade route.

In 1854 Heinrich Barth, a German, navigated the headwaters of the river Benue and plotted the map of the Middle Niger, while his friend, Adolf Overweg, surveyed Lake Chad. In 1873, Luis Eugéne Dupont and Oskar Lenz discovered the river Ogooué. The way was opened up for hydrological survey. It took another thirty-four years for the mean annual discharge of the Niger River at Koulikoro (39 793 cfs) to be estimated for the first time.

1.4 CONGO BASIN

In 1483, Diego Cão, sailing along the coast of West Africa, discovered a stream of fresh water of a different color from the sea, thirty kilometres off the coast. He became the first European to land on the bank of the Congo.

Three centuries later, Mungo Park thought that the stream was the Niger. In 1812, the British Government sent Captain James Kingston Tuckey to explore the river. Just as on the Niger, fever killed several members of the expedition and it failed completely. In 1866, Livingstone started his expedition to find the sources of the Nile and Congo, hoping to discover that Herodotus's fountains were the source of both streams. Later, when navigating the Luapula river system, he believed that Lake Mweru contributed to the Nile sources.

In 1877, Henry Morton Stanley's expedition arrived at Nyangwe after a 999-day trip through the Congo basin. Of the 350 men, women and children, only 115 reached the final point. Fever, smalpox, battles with the hostile tribes, drownings and crocodiles took this heavy toll. Only a few years later, was it proved by Thompson that Lake Tanganyika, through the Lukuga outlet, also belonged to the Congo system.

River Kasai, a tributary of the Congo, was put on the map by Wissman and many southern tributaries were explored and mapped by Grenfell. The discovery of the Ogooué and Congo basins was more or less a matter of competition between the French and Belgian Governments. Pierre Savorgnan de Brazza established a post on the Ogooué in 1880 and another post, Brazzaville, on the Congo. Then Stanley reached the Congo for the second time, while scouting for the Belgian king, and he established Léopoldville, today Kinshasa, opposite to Brazzaville.

Meanwhile, Savorgnan de Brazza extended his travels to the regions beyond the confluence of the Ubangi and Congo rivers and Paul Crampel explored the Sangha River basin. In 1891, he was killed while exploring the Chari river in an attempt to extend French rule to Lake Chad.

Because the river has so long been a boundary between two countries, hydrological problems were never of great importance and existing records from the beginning of the twentieth century are of poorer quality than those of the other main African streams. (See Fig. 1.2 — enclosure.)

1.5 THE ZAMBEZI IN AFRICAN HISTORY

It was during the Pleistocene that the Zambezi river created the Batoka Gorge and the famous Victoria Falls on the boundary between the low floor of the gorges and higher basaltic uppermost-Tertiary surface. Gravel and silt deposited during the flood periods in the flood plains of the Middle Zambezi formed areas of human settlement. The plains are still subject to seasonal flooding, with an annual rise and fall of the water level and a corresponding human and ecological response. The increasingly important part played by the Zambezi valley in African history has already been traced back to the first millenium A. D. External contacts and foreign commercial interests played a vital role in the economic development of the Early Iron Age. Some early Iron Age localities, such as Dambwa near Livingstone, indicate possible contacts between Early Iron Age and Late Stone Age cultures. The cultural and economic centres spread out into the valleys of the coastal tributaries such as the Kafue river. The routes of the traders from the Indian Ocean followed the great rivers flowing from the interior, such as the Sabi and the Zambezi. Once the African plateau was reached through the escarpments, there was easy access to the mining areas of iron, copper and gold. The Middle Zambezi Valley flood plains have also always been famous for their large herds of game, particularly elephants and thus ivory, as well as ores were among the important commodities of foreign trade.

The tribe of the Tongas has been settled in the Zambezi valley at least since 1100 A. D. and is one of few African tribes whose origin can be traced back to the Early Iron Age.

A further increase in the importance ot the Zambezi river came with the extension of commercial contacts in the fifteenth century. In 1514—1516 Antonio Fernandes followed the Sabi and Zambezi while looking for new routes into the African interior. Since then, numerous deviations have led from the main route in the Zambezi valley and have frequently followed its tributaries. A result of the Portuguese extension of the main trade routes there is good knowledge of the interior hydrography. The river Zambezi is plotted as the Cuana or Zambere river on the 1700 A. D. map of F. Morden. Lake Zaire can be identified as today's Tanganyika and River Zambere as to Luapula.

During the seventeenth century, the Portuguese tried to establish a trans-African route along the Zambezi connecting Angola and Mozambique. In 1798—99, Dr. L. Lacerda deviated from the Zambezi at Tete to Lake Mweru and to Kazembe's court; he died there in 1799. Captain Gamitto explored the Luangwa valley in 1831, and in 1853—54 J. da Silva reached Lealui on the Upper Zambezi, then already a centre of the Lozi Kingdom.

The history of the Victora Falls and the Zambezi river is connected with Livingstone's explorations. He visited the Falls during his first trip in 1855. He started his journey in 1851 from the south. After reaching River Chobe, Livingstone followed the Zambezi north, then turned in the direction of Luanda. On the way back, he canoed the river from Sesheke down to one of the small islands at the edge and he saw "… the columns of water appropriately called smoke, rising at a distance of five or six miles, exactly as when large tracts of grass are burned in Africa."

European explorers are not the only people to be recalled in Zambezi history. The Barotse tribe, originating from Lozis, has been settled along the Zambezi flood plains for several centuries. This tribe can serve as an excellent example of a people living in a natural balance with the water environment up to the present time. Water plays a significant part in national customs. The Barotse Kings established their capital in the centre of the Zambezi flood plains and the royal seat is surrounded by water for almost half of the year. Once the water level has reached a certain stage, a pageant called "Kuomboka" is held and the King and his court leave the summer palace in the royal barge for the winter capital at Limulunga to rule from here until the river once again drops below the critical level.

In 1886, a Czech explorer, Dr. Emil Holub, [13] operated north of the flood plains of the Zambezi tributary, the river Kafue and there carried out the very first hydrological measurements. Twenty years later, on the same river, the first hydrographic observation post was opened at the Kafue Bridge connecting South Africa with the Central African Interior. Thus, the first base for hydrological survey in tropical Africa was established. An exciting period of hydrographical explorations in the African tropics was at an end and a new road opened up for scientific hydrology.

1.6 LIST OF LITERATURE

[1] Anděl, J., Balek, J., Verner, M., 1971. An analysis of the historical sequences of the Nile maxima and minima. Symposium on the role of hydrology in the Econom. Dev. of Africa. WMO No 301 Rep., Addis Abbaba, pp. 24—36.

[2] Barguet, P., 1953. La stéle de la famine á Sehel. IFAO Bibliotheque d'etude t XXIV. Le Caire, pp. 15.

[3] Biswas, A. K., 1970. History of hydrology. North Holland Pub. Comp.

[4] Borchard, C., 1906. Nilmesse und Nilstandsmarken, Berlin.

[5] Cooke, H. B. S., 1946. Observations relating to Quaternary environments in East and Southern Africa. Geol. Soc. South Af. Bull, Vol. 20, 73 p.

[6] Clark, J. D., 1963. Ecology and culture in the African Pleistocene. S. Afr. *Journal of Science* IX, No. 7 pp. 363—66.

[7] Fogan, B. M., 1968. Short history of Zambia. Oxford Univ. Press, Nairobi, 165 p.

[8] Flint, R. F., 1959. Pleistocene Climates in Eastern and Southern Rhodesia. Bull. of the Geog. Soc. of America, Vol. XX, 343—74.

[9] Drioton, E., 1953. Les origines pharaoniques du Nilometre de Rodah. Bull. de l'Institut d'Egypte, t. XXXIV, Le Caire, pp. 291—316.

[10] Fordham, P., 1965. The Geography of African affairs. Penguin Books, London.

[11] Gamitto, A. C. P., 1960. King Kazembe. Centro de Ciencias Politicas y Socias, Lisboa.

[12] Gardiner, A. H., 1947. Ancient Egyptian Onomastica. Oxford, Vol. II, pp. 139.

[13] Holub, E., 1890. Von der Kapstadt ins Land der Maschukulumbe. Hölder, Wien.

[14] Hurst, H. E., 1964. Le Nil. Le Payot, Paris.

[15] Leakey, L. S. B., 1964. Prehistoric man in the tropical environment. Symposium on the Ecology of Man in the Trop. Environ., Nairobi. Morges, pp. 24—29.

[16] Legrain, G., 1896. Textes gravées sur le quai de Karnak. ZAS 34, pp. 111—118.

[17] McDonald, J. F., 1955. Zambezi River. MacMillan, London 1955.

[18] Moorehead, A., 1960. The White Nile. Hamish Hamilton.

[19] Moorehead, A., 1962. The Blue Nile. Dell Publish. Comp., New York.

[20] Nilsson, E., 1932. Quaternary glaciations and pluvial lakes in British East Africa. Geogr. Ann., Stockholm, Vol. 13, pp. 249—349.

[21] Olivier, R., Fage, J. D., 1962. A short history of Africa. Penguin Books, London.

[22] Popper, W., 1951. The Cairo Nilometer. Univ. of Calif. Publ. in Sem. Phil., Vol. 12, Berkeley.

[23] Reisner, G. A., 1960. Senna Kumma. Boston 1960.

[24] Sethe, K., 1905. Beiträge zur ältesten Geschichte Ägyptens. Leipzig, 105 p.

[25] Summers, D. F. H., 1960. Environment and culture in Southern Rhodesia. Proc. Amer. Phil. Soc., 104, pp. 266—99.

[26] Turnbull, C. M., 1964. Forest hunters and gatherers: The Mbuti pygmies. Ecology of Man in Tropic. En., Nairobi. Morges, pp. 38—43.

[27] Toussoun, O., 1925. Mémoires sur l'historice du Nil, t. 2/MIE, VIII, pp. 265—266.

[28] Verner. M., 1972. Periodical Water Volume of the Nile. Ar. Orient., Prague, 40 p.

[29] Wayland, E. J., 1934. Rifts, rivers, rain and early man in Uganda. London Royal, *Ant. Inst. Jour.*, Vol. 64, pp. 333—352.

2. CLIMATOLOGY OF THE AFRICAN TROPICS

2.1 EFFECT OF THE SUN ON TEMPERATURE, AIR PRESSURE AND AIR MASS MOVEMENT

The tropic regions are bound by two parallels of latitude, 23°27′ north and south of equator, which are also called the Tropic of Cancer and the Tropic of Capricorn. Meteorologists sometime take the tropics to be the area between the Easterlies and the Westerlies. The symmetry of the tradewind belts is distorted in the tropics by the presence of low pressure formed by the greater heating of the land than of the sea surface, by the shifting of the low pressure over Africa and the shifting of the belts of subtropical heights.

The sun culminates between Cancer and Capricorn within the angles 43° − 90° and the duration of a day is 10.75 − 13.25 hours. The sun is overhead in the tropics twice a year and the temperature is under the influence of the sun. At the equator, the sun is overhead on March 21 and September 22 and the solar radiation is highest at these points. When the sun is in the solstices, the intensity of its radiation is at the minimum, nevertheless the difference between minimum and maximum is much lower then in the middle latitudes. The variability of mean annual and monthly temperatures increases southward and northward of the equatorial region.

The heating of the earth's surface is greater in the equatorial region than outside of it. Solar radiation on a hypothetical horizontal surface at the top of the atmosphere

Tab. 2.1. The values of solar radiation in $Kcal/cm^2/day$ at various latitudes

30 S	20 S	10 S	0	10 N	20 N	30 N	Month
1.016	0.989	0.936	0.859	0.758	0.638	0.479	J
0.930	0.943	0.929	0.887	0.821	0.729	0.620	F
0.799	0.858	0.891	0.895	0.875	0.824	0.753	M
0.642	0.740	0.818	0.869	0.896	0.897	0.870	A
0.509	0.632	0.737	0.823	0.887	0.928	0.947	M
0.442	0.573	0.692	0.793	0.875	0.935	0.974	J
0.469	0.595	0.709	0.803	0.878	0.930	0.957	J
0.573	0.682	0.774	0.842	0.885	0.906	0.898	A
0.726	0.802	0.852	0.879	0.815	0.848	0.794	S
0.875	0.904	0.907	0.885	0.835	0.767	0.663	O
0.986	0.969	0.927	0.862	0.774	0.665	0.538	N
1.034	0.996	0.932	0.846	0.737	0.611	0.465	D

is highest in the centre of the tropics. Tab. 2.1 gives the values of solar radiation in cal/sqcm/day for the tropics and adjacent subtropics, these values being a significant factor in various water-balance studies.

Fig. 2.1 gives a comparison between solar radiation at the top of the atmosphere and that at the earth's surface.

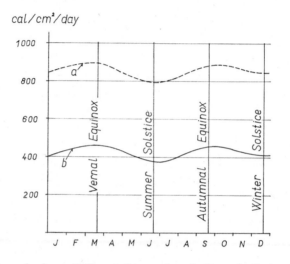

Fig. 2.1. Fluctuation of solar radiation at the equator, a) above the limits of the atmosphere, b) received at the earth surface.

Another factor particularly significant in evapotranspirational studies is the maximum possible hours of sunshine. Table 2.2 provides information on these values in the tropics and outside their boundaries.

Tab. 2.2. Maximum possible hours of sunshine at selected latitudes

30 S	20 S	10 S	0	10 N	20 N	30 N	Month
13.9	13.2	12.6	12.1	11.6	11.1	10.4	J
13.2	12.8	12.5	12.1	11.8	11.5	11.1	F
12.4	12.3	12.2	12.1	12.1	12.0	12.0	M
11.5	11.2	11.9	12.1	12.3	12.6	12.9	A
10.7	11.2	11.7	12.1	12.6	13.1	13.6	M
10.3	10.9	11.6	12.1	12.7	13.3	14.4	J
10.4	11.1	11.6	12.1	12.6	13.2	13.9	J
11.1	11.5	11.8	12.1	12.5	12.8	13.2	A
12.0	12.0	12.1	12.1	12.2	12.3	12.4	S
12.9	12.6	12.3	12.1	11.9	11.7	11.5	O
13.6	13.1	12.6	12.1	11.7	11.2	10.7	N
14.0	13.3	12.0	12.1	11.6	10.9	10.3	D

The actual number of hours of sunshine (apart from the deserts), is much lower in the African tropics owing to the extensive cloudiness. During the hot season the clouds are of a cumulus type produced by the convection currents and layer clouds, formed by the spreading out of cumulus and cumulonimbus, are common as well. In the vicinity of low pressure, continuous sheets of rainclouds (nimbostratus) occur. A similar type of cloud is also formed after the invasion of maritime air and, under the influence of orography, these clouds thicken. Turbulence clouds occur in the early morning when the ground is wet. On the Central Plateau, the fine weather is occasionally interrupted by so called gutti spells, which are periods of overcast, drizzly weather associated with the invasion of maritime air.

Subtropical belts of high pressure have a dominant influence on the climates of Africa. Local topography is less important then on other continents because the greater part of the African tropics consists of a high plateau, rising more than 1000 meters above sea level. There are not as many mountains chains influencing the climate as in tropical America and Asia. The seasonal system of winds and rainfall therefore depends mainly on the broad atmospheric circulation pattern with a few local deviations.

Following the overhead position of the sun, the zone of greatest heat shifts from the southern hemisphere in January to the northern in July, followed by a pressure trough with a lag of about one month. Moisture is brought to the continent through the inflow of maritime air in the form of air streams associated with the subtropical heights over the ocean. Rainfall occurs either as orographic or convective, depending on the temperature, location and season.

There is a simplified scheme of the air circulation over tropical Africa in July and January in Fig. 2.2. Low pressure located close to the equator in July and January merges with the monsoon situated near latitude 20° S. Strong streams are dominant. Along the west coast, south of the mouth of the Congo, the southeast trade wind is directed parallel to the coast, raising a southerly and offshore wind. Along the northeast coast, an Asian northeast monsoon, mixed with the trade winds, is drawn south across the equator. Another airstream associated with the flow from the northern subtropics penetrates as far as the Capricorn. By July, the low pressure and inter-tropical front, also called the Intertropical Convergence Zone — ITCZ, migrates north of the equator. The Southern subtropical high pressure is near Capricorn and the westerlies, with travelling cyclones, affect only the southern part of the continent. With the sun moving in a northwards direction, the interior of southern Africa becomes cooler than the Indian Ocean and a high-pressure system is developed. The descending and outflowing air causes a dry season. North of the equator, the east coast escapes the high-pressure system and, under the influence of trade winds, rainfall is caused by the air masses flowing from the Indian Ocean.

Obviously, the activity of the sun and its influence on the hydrological phenomena is more pronounced in the tropics than elsewhere.

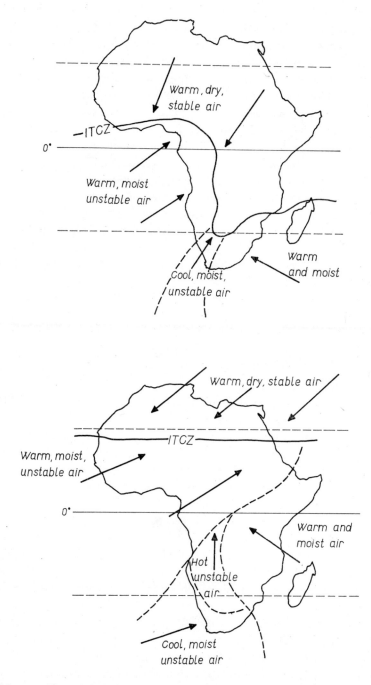

Fig. 2.2. Mean position of the Intertropical Convergence Zone, a) in January, b) in July.

2.2 TROPICAL STORMS AND RAINFALL

A general requirement for precipitation is the lifting and cooling of large volumes of air. Beside orographical conditions, storms are the main centers making the air rise and thus producing rainfall. A large quantity of moist air is another requirement for the formation of precipitation, not only because of the water itself but also because of the large amount of latent energy in the water vapor required for the formation of storms.

There is a great variety of storms in the tropics, but only a few types are characteristic for Africa. The generally used term "cyclone" can be applied to a great variety of storms. Cyclones typical for the intermediate zones occur in the tropics of Africa on a rather small but more intense scale and only east of Madagascar. They originate over the ocean in fact, and have little resistance to the wind. Trewartha [19] lists the following typical features of tropical cyclones:

a) The isobars are more symmetrical and more nearly circular,
b) pressure gradients are steeper and winds are stronger (over 120 km/hour),
c) rains tend to be torrential and more evenly distributed along the center,
d) they are more numerous during the warm than the cold season,
e) they have no anticyclonic companion.

The weaker tropical lows are frequent sources of African tropical rainfall and they are mainly productive of benefical rainfall as opposed to convectional showers resulting from surface heating. The rainfall from weak tropical lows is much more extensive, not very vigorous, but of longer duration and it falls from skies which are uniformly overcast. The lows appear as wave-like disturbancies along the tropical fronts and many of them appear as shallow depressions associated with showery rain.

Thunderstorms are local storms accompanied by thunder and lightning which are not the causes of the storms. The potential energy of the latent heat of condensation and fusion in the moist air, conditionally or convectively unstable, is rapidly converted into the kinetic energy of violent vertical air currents associated with torrential rain, hail, lightning and thunder. The rapid vertical movement is associated with high surface temperatures and vigorous convectional overturning. Thus, thunderstorms occur most frequently in the warmer latitudes of the earth, in the warmer seasons and in the warmer hours of a day. Humid heat and thunderstorms are closely related. According to various sources, the relative humidity of the air masses must exceed 75%.

Rainfall is the most significant climatic factor affecting the economy of tropical Africa. It waters the crops during the rainy seasons, replenishes the lakes, swamps, groundwater storage, and makes irrigation possible during the dry seasons. It is vital to all aspects of tropical economy and a detailed examination of the rainfall regime is essential in all phases of planning and development.

The distribution of rainfall in tropical Africa in time and area is very uneven, except for the equatorial regions. A great part of the tropical territory enjoys rainfall for less than four months of the year. Annual totals are closely associated with the classification of African climatic regions. In a simplified scheme Barry [6] recognizes

    a) equatorial regime,
    b) tropical regime.

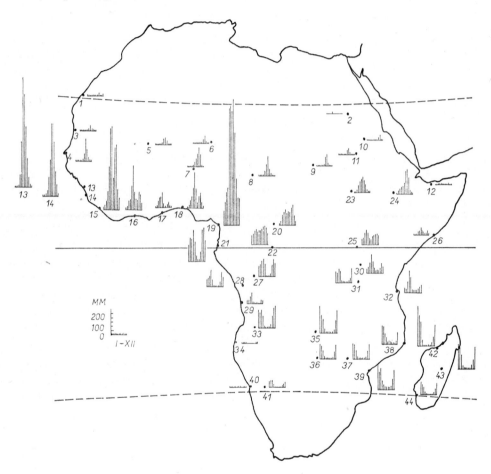

Fig. 2.3. Monthly distribution of rainfall in the African tropics.

Here, equatorial rainfall is characterized as occurring throughout the year with two maxima annually and small variability from year to year. Rainfall is associated with equatorial low pressure, although rainy periods result from disturbances. The tropical regime has summer rainfall alternating with a dry winter and an annual amount of rainfall between 250 and 1000 mm. As will be shown later, such a clas-

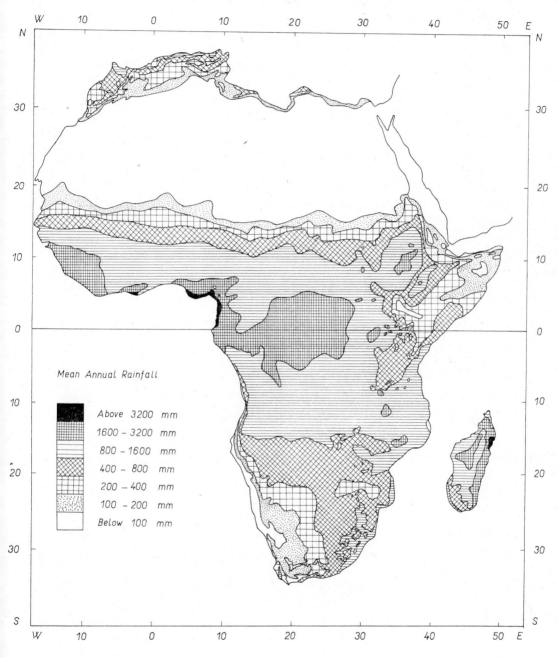

Fig. 2.4. Mean annual rainfall in tropical Africa.

sification can only be taken as very approximate. A rather low density of rainfall gauges in Africa has not allowed of more extensive studies, although a considerable amount of work has already been done. Lebedev [11] based his African climatological study on the analysis of 1564 rainfall gauge data; 50% of them have been recording for more than 20 years and 23% for more than 30 years. Several reliable records date from the beginning of the century.

The annual amount of rainfall in Africa lies anywhere between 0 and 15 000 mm, monthly totals reaching up to 3000 mm. A value of at least 1 mm is usually given as a long-term annual mean elsewhere in the tropics. A simple map provides information on the monthly distribution of rainfall in tropical Africa (Fig. 2.3). A map in Fig. 2.4 gives a picture of the distribution of annual rainfall in the tropics. The rainfall belts as well as the association with the ITCZ can be seen on both maps. The influence of orography can be traced on the maps – the rainfall distribution is different in the southern and northern tropics. In the northern hemisphere a zonal distribution of rainfall can be traced between 11° and 18°.

Tropical rainfall stems largely from cumulus clouds and thus heavy rainfall is more frequent in the tropics then in the middle latitudes. This fact is particulary

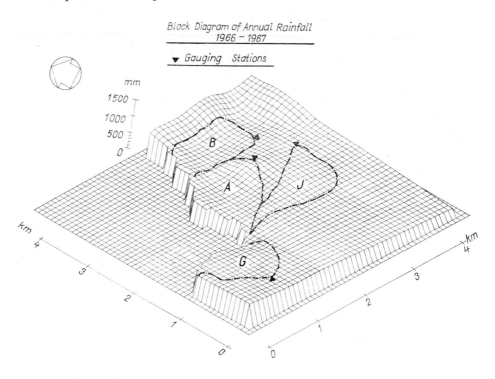

Fig. 2.5. Fluctuation of the annual rainfall at the Luano experimental catchments, 12°34'S, 28°01'E, 1324 metres.

significant in the diurnal rhythm when diurnal heating leads to an afternoon maximum of precipitation as a result of convective downpours. In such a vast area as are the tropics, different patterns are also common. Khartoum in the Sudan can serve as an example; it receives 80% of its rainfall between 18.00 and 06.00 hours, because of the afternoon thunderstorms brought by upper winds from the hills bordering the Red Sea.

As was proved experimentally [5] the distribution of annual rainfall is very uneven, even within small areas. In Fig. 2.5 a map indicates the fluctuation of annual rainfall within an area of 16 km^2 on the Central African Plateau. This fact may be not found so surprising for single storm events as for monthly and annual totals.

Lebedev [11] analysed the probabilities of occurrence of annual rainfall. The following are the values for selected stations in the tropics (in mm):

Tab. 2.3. Probability of occurrence of annual rainfall (mm) at some stations in the African tropics. After Lebedev

Station	Annual mean rainfall	Probability of occurrence						
	mm	5%	10%	20%	50%	80%	90%	95%
Khartoum	158	332	278	222	142	87	69	57
Freetown	3655	4730	4410	4020	3520	3000	3190	2740
Addis Ababa	1266	1750	1695	1455	1215	1045	990	960
Livingstone	722	1040	980	872	720	564	488	430
Windhoek	372	690	610	475	330	255	210	175

The values given in Tab. 2.3 in terms of percentage of the mean annual rainfall provide only a very approximate estimation of the probability of occurrence of each particular region where only annual totals are known.

There are not enough data available for studies of the frequency and intensity of the rainfall pattern, except for experimental catchments. In most cases, data based on daily totals are available and for a more advanced analysis autographic recorders must be installed in advance.

A widespread characteristic of tropical rainfall is the concentration of annual rainfall into a relatively small number of days. Barry [6] estimates that about half of the annual precipitation in the Kenya part of the Rift Valley falls within only 13% of the rainy days. Similarly, in Uganda, storms with a total of 25 mm form 30% of the annual total and the peak rates of some rainstorms observed were between 250 and 350 mm per hour. Information on the return periods of maximum daily rainfall in some locations of the Central African Plateau has been given by Bailey [2]. Tab. 2.4 contains the return periods for 4 stations within the boundaries of Zambia, calculated in inches.

Tab. 2.4. Return periods of daily rainfall (inches) for some stations of the Central African Plateau. After Bailey

Station	Location		Return periods in years					
			2	5	10	25	50	100
Kasama	10°10′ S	31°12′ E	2.63	3.21	3.60	4.08	4.44	4.80
Livingstone	17°48′ S	25°58′ E	2.52	3.52	4.16	4.95	5.56	6.15
Lundazi	12°30′ S	33°20′ E	2.21	3.50	4.02	4.68	5.16	5.65
Lusaka	15°03′ S	28°30′ E	2.28	2.81	3.16	3.62	3.95	4.28

The analysis was made by the Lieblein procedure using Fisher-Tippett Type I distribution, which is found to be very suitable for tropical rainfall. Another series of interesting values on the probability of daily rainfall equal to or greater than a selected amount were obtained in a similar manner (Tab. 2.5).

Tab. 2.5. Probability of occurrence of daily rainfall for some stations of the Central African Plateau. After Bailey

Station	Location		Probability of daily rainfall equal to or greater than					
			2	3	4	5	6	inches
Kasama	10°10′ S	31°12′ E	91	28	4.7	1	1	%
Livingstone	17°48′ S	25°58′ E	76	33	11	3.4	1	%
Lundazi	12°30′ S	33°20′ E	85	37	10	2.5	1	%
Lusaka	15°03′ S	28°30′ E	71	12.5	1.8	1	1	%

A similar analysis was attempted for some regions of the African tropics, based on annual totals. Lebedev calculated annual rainfall totals of certain probability of occurrence in the regions of savanna (Fig. 2.6).

In the absence of autographic records, information on daily maximum rainfall can be useful in flood-regime studies, particulary, when considering that in most cases the duration of storms does not exceed 8 − 10 hours (Tab. 2.6).

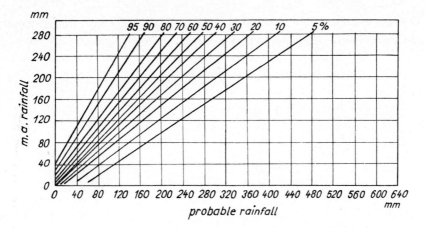

Fig. 2.6. Relationship between mean annual rainfall and annual rainfall of certain probability of occurrence in the tropical savanna. After Lebedev.

Tab. 2.6. Maximum daily rainfall (mm) for selected tropical stations

Country	Maximum daily rainfall (mm) in											
	J	F	M	A	M	J	J	A	S	O	N	D
Mali	17	111	73	82	135	111	140	200	170	90	48	14
Chad	0	21	58	81	88	100	126	156	128	78	28	0.2
Sudan	82	114	93	129	162	150	285	171	158	128	162	110
Senegal	30	28	27	29	72	152	167	242	238	119	84	6
Siera Leone	69	38	64	64	102	102	190	213	229	277	89	86
Ghana	89	107	112	140	295	429	94	81	114	168	104	76
Nigeria	117	124	122	152	168	226	190	193	315	190	196	94
Central Af. Republic	71	117	135	150	165	164	154	144	142	216	112	89
Ethiopia	67	60	74	104	94	130	94	127	112	71	69	157
Somalia	74	38	132	112	91	155	38	53	53	53	79	33
Kenya	147	127	96	127	206	142	142	145	112	152	119	206
Tanzania	127	124	146	203	206	127	142	71	130	201	201	117
Zaire	165	129	159	175	150	142	124	172	178	149	147	198
Zambia	170	138	203	124	38	18	8	30	33	99	127	119
Angola	164	265	225	158	112	42	39	52	78	144	135	150
Mozambique	414	328	386	358	124	168	99	94	112	107	259	221
Botswana	130	117	170	61	34	67	17	24	72	83	61	110
Malawi	104	96	130	81	28	23	13	23	23	58	69	145
S. Rhodesia	170	104	115	79	41	41	18	38	53	79	132	112
Namibia	127	123	84	121	58	21	10	20	33	34	69	84

The calculation of depth-area-duration curves frequently accompanying the hydrometeorological studies is still very rare in Africa, except in some areas where special projects have been completed. For instance, in the course of the WMO survey of the Upper Nile basin, a density of one rainfall station per 350 km^2 was achieved, with 15% of the stations being autographic recorders. Depth-area-duration curves were calculated here for one or two single storms. The depth of the areal rainfall in the Nyando basin, part of the Kenyan tributaries of Lake Victoria, was calculated by Raman [15]. The values in Tab. 2.7 are given in mm.

Tab. 2.7. Distribution of the rainfall (mm) in Nyando basin, Kenya

Date of storm	Area (km^2)				
	10	100	1000	2000	5000
27/11/61	83.0	80.0	65.4	53.0	34.0
27/4/65	58.4	46.0	35.0	30.0	22.0

For the purpose of the depth-area-duration analysis within the area, the following formula was developed:

$$I_t = 26.84 \, (t + 0.4)^{-1}$$

where I_t is the intensity in mm/hour and t is the duration of the storm in hours. It was calculated that for the duration of 3 hours, the intensity is 7.9 mm/hour and a total of 23.7 mm, for 6 hours, it is 4.2 mm and 25 mm, for 12 hours 2.2 mm/hour and 26.4 mm and for 24 hours, 1.1 mm/hour and 26.4 mm. Obviously, this type of value cannot be extrapolated far from the limits of the studied area, however, useful information is given on the relatively low variability of rainfall totals from storms of various duration.

2.3 EVAPORATION AND EVAPOTRANSPIRATION

Next to rainfall, evaporation and evapotranspiration, both under the direct influence of sunshine, are the most significant phenomena in the hydrological cycle of the African tropics. Nevertheless, more advanced studies of evaporation/evapotranspiration, based on energetic budgeting, have been initiated in tropical Africa only very recently. Evapotranspiration measurements in particular, require further extension, since the role of tropical plants in the water balance should be emphasised.

Penman's method of calculation of potential evaporation has become very popular and is widely used in Africa in combination with direct measurements. Application of the standard Weather Bureau Class A pan under tropical conditions indicated that under direct sunshine the metallic pan becomes much hotter then the water it contains and thus the measurements are influenced to a great extent. Several adaptations in design have therefore been proposed by hydrologists working under tropical conditions. Any adaptation, however, detracts from the standardization. Another problem, when working with the evaporation pan in the tropics, is protection against animals and birds. A standard tropical pan was recommended back in 1959 by organisations concerned in Kenya, Uganda, Tanzanie, Zambia, Southern Rhodesia and Malawi, which is screened by thin wire netting and coated inside with black bituminous paint. Another design used was a fibreglass pan either screened or unscreened. By using fibreglass, the influence of the sun on the heating and cooling of the material has been reduced. The data obtained from both pans differ significantly as can be seen from Tab. 2.8.

Tab. 2.8. Observed evaporation from galvanized and fibreglass pans at the Luano catchments (12°34′ S, 28°01′ E)

Month	Evaporation from galvanized Class A pan (mm)	Evaporation from fibreglass pan (mm)
October	227.48	170.43
November	135.94	120.20
December	129.89	106.44
January	111.14	99.34
February	127.85	103.18
March	127.40	115.14
April	136.68	127.68
May	138.45	122.62
June	119.96	102.28
July	135.06	116.99
August	173.45	155.03
September	193.65	162.77
Total 1967/68	1756.95	1502.15

The annual totals as observed at the boundaries of Zaire and Zambia differ by almost 17%. Another difference occurs through different interpretation of Penman's formula. In the tropics of Africa numerical methods of the calculation as described by McCulloch and by Aune [1] have been extensively used; both are equally correct, Aune's method is better applicable when a computer is to be used. The following difference has been found at the same location as that described in the previous Table (Tab. 2.9):

Tab. 2.9. Potential evaporation calculated from Penman's equation by McCulloch's and Aune's methods for the area of the Luano catchments (12°34' S, 28°01' E)

Month	Aune's method (mm)	McCulloch's method (mm)
October	184.1	187.5
November	166.2	170.5
December	147.2	149.4
January	142.5	145.3
February	143.0	151.4
March	162.3	169.3
April	146.4	154.0
May	124.5	136.8
June	101.6	114.6
July	115.1	126.8
August	136.2	146.6
September	161.6	168.7
Total in 1967/68	1730.2	1820.9

A difference of more then 5%, as in this case, may cause serious difficulties in the water-balance calculations, but it should always be taken into account.

Another problem is foreseen in the collection of the data needed for the interpretation of Penman's or any other formula. While the air temperature can normally be readily obtained and the dew point temperature calculated, wind speed and radiation are factors rarely available at many localities. The wind factor can be ignored in approximate calculations; radiation is frequently replaced by hours of bright sunshine. As has been proved experimentally [4], the data obtained by using sunshine and by radiation differ significantly (Tab. 2.10).

A comparative calculation based on both values is recommended for water-balance studies, otherwise a comparison made with results obtained only by applying one factor is not reliable. The difference between the calculated and observed values should also be considered. According to the results of comparative measurements and calculations, the deviation varies month by month and year by year and no reliable coefficients for any transfer of the values can be determined.

More detailed evaporation studies are only available for some regions. For instance, in Uganda Rijks set up monthly and annual maps of potential evaporation which can serve as a guidance when similar data are prepared. The highest values of annual evaporation, are 2000 mm at the border between Uganda and the Sudan. The Water Management Department of Zambia measured 2492 mm at Samfyia,

Tab. 2.10. A comparison of the calculated potential evaporation by sunshine and radiation (Aune's method).
Luano catchments (12°34′ S, 28°01′ E)

Month	Aune's method applied by using	
	sunshine (mm)	radiation (mm)
October	178.64	156.23
November	188.86	173.72
December	155.60	146.99
January	163.41	150.48
February	140.08	130.82
March	176.00	156.85
April	149.66	134.29
May	131.90	113.60
June	108.67	90.12
July	118.56	99.76
August	137.89	114.60
September	168.91	142.17
Total 1969/70	1818.18	1609.13

Bangweulu Swamps in 1957/58. The mean evaporation at Merowe, between Khartoum and Wadi Halfa, is given as 8,4 mm per day which gives an annual value of 3066 mm; another source estimates that the annual evaporation exceeds 4000 mm in arid regions of Sudan.

2.4 CLIMATE OF TROPICAL AFRICA

Tropical Africa does not experience the great temperature contrasts of Europe, Asia and North America, except for some parts of the tropics. There is a great variability of climate in the African tropics, mainly owing to the movement of air masses, differing in air moisture and relative stability rather than in temperature. The air masses get into contact with each other along the Intertropical Convergence Zone which moves across the continent as a response to the winds and temperatures. Sometimes it is described as an equatorial trough of low pressure. The converging northeast and southeast trade winds are among the main factors forming the tropical climate.

A climate is not something which can be unambiguously defined and thus any attempt to classify the typical climatological regions is strongly subjective. Many attempts have been made to do so. It may be easier to make a classification in the tropics than elsewhere, although the views of authors differ greatly. The following survey of classifications, which may be found incomplete by a climatologist, has been made solely for the purpose of hydrology. At this point, mention may be made of a classical statement by Voeychkov [20] who considered rivers to be pure products of the climate.

The classification of Herbetson [9] is an historical example. He differentiated the tropics as:

a) Tropical deserts of the Sahara type,

b) Intertropical tablelands of the Sudan type,

c) Equatorial lowlands of the Amazon type.

An equally simple type of classification was formed by Gourou, distinguishing between

a) Tropical zones, where precipitation exceeds 800 mm

b) Desert zones, where precipitation is below 800 mm.

Fosberg, Garnier and Krichler [7] defined the humid tropical region as having a mean monthly temperature of at least 20 °C for eight months of the year, vapour pressure of at least 20 milibars, relative humidity of 65% for six months, a mean annual rainfall of more than 1000 mm and a mean monthly total of at least 75 mm for six months of the year. By applying these values, a wide section of the tropics can be considered as a single region. Since the regimes of tropical African rivers vary to a much greater extent, a more detailed classification is needed for the purpose of hydrology. From this point of view, the classification of Pollock [3] is acceptable; this differentiates between:

The equatorial type found on a narrow strip of the West African coast and in the northern Congo basin. The basic characteristics are continuous heat, high humidity and well-distributed rainfall. Mean monthly temperatures vary between 24 – 27 °C, with only slight seasonal variation. The seasonal variation is less than 3 °C, the diurnal variation may reach up to 8 °C. Temperatures, however, are not as high in as the arid regions. Other general characteristics are the lack of air movement and high humidity.

The modified equatorical type under the influence of monsoonal winds in rather narrow strips of the West African coast and the eastern part of Madagascar.

A tropical continental climate practically surrounds the equatorial regions and is characterized by seasonality of rainfall and increasing variation in temperature. Rainfall varies considerably between 500 and 1500 mm, the lowest being at the edge of the steppe region and the highest near the equatorial forest. There are two typical seasons, one wet and hot during and after the period of high sun and the cool season, noticeably dry, when the sun is in the opposite hemisphere. The dry season becomes more pronounced with the distance from the equator, so that at the margins of the zone, it lasts eight months. The temperature increases from the equator to the desert margin.

The high plateau *modifies the tropical continental climate* so that the climate is cooler. Thus, the winter is the most pleasant period of the year with many sunny days and cool nights. The zone is characterized by sharply marked seasons, a high rate of precipitation and the occurrence of frost. The tropical lowland and highland regions are in both the Southern and Northern Hemisphere and in Malagasy.

The semiarid type of climate sometime called the tropical steppe climate, is found near the desert margins of the Sahara in a rather narrow belt, in Southern Kalahari and in East Africa as a narrow belt far from the sea. The region is characterized by high temperature, and 250 – 500 mm annual rainfall which is extremely variable in amount and distribution, mostly taking the form of thunderstorms.

The climate of the deserts is characterized by negligible rainfall, some areas being without precipitation for one or more years. Other characteristics are extreme temperature variability with the maximum above 50 °C and the minimum below freezing point, low humidity, sunshine at maximum (Sahara has a maximum of 90%). Due to the shape of the continent, there are more desert areas in the Northern African Hemisphere, while the south is under the influence of the oceans.

The climate of tropical highlands. These are the regions locally significant in Ethiopia, Uganda, Tanzania and Kenya. Parts of South Africa also have a climate influenced by high altitude. Frost may occur three to six months a year, rain is generally over 500 mm, but decreases rapidly in the rain shadow areas. Stamp [17], using a slightly different type of classification (Fig. 2.7 see enclosure), characterizes the *equatorial climate* by monotonous constant heat, humidity and rainfall, very little seasonal rhythm, but no great extremes. Some stations record two rainfall maxima per year, some, toward the margins, have a season with more rain then the rest of the year. Typically there is no dry season, which can be considered as the main distinction from other regions. In many places, in the absence of wind, the rain is convectional and thus falls in the same area from which the evapotranspiration originated. The windward coast of Madagascar, Mauritius and Réunion, under the influence of the southeast trades, are taken as a separate subregion, where the dry period is more pronounced, however there is at least 3 mm of rainfall a month. *The tropical savanna climate* is defined as Africa's most typical climate. It covers a greatly variable region between the equatorial forest and the desert margins with a precipitation varying between 2000 and 400 mm. The rhythm of the seasons is influenced by the rhythm of the temperature. A cool dry season alternates with a hot dry season later on producing rains which lower the temperature; however, in areas where there is too little rain, the temperature reduction is not so pronounced. Rainfall totals vary considerably from year to year because the rain mostly takes the form of heavy showers and thunderstorms. The tropical savanna climate is bounded by the climate of *low latitude steppes and the low latitude deserts* by the 400 mm isohyet. The rains come very irregularly in the form of storms and there is

38

no pronounced rainy season. Temperatures are extreme, with a recorded maximum of 56.6 °C in the shade and 76.8 °C at ground level, and frosts are usual. There is very little humidity in the air and so the human body perspires up to 6 litres a day. The conditions south of the equator are less severe due to the influence of the oceans. The southern deserts are therefore sometimes considered to be semideserts. *Undifferentiated highland climate,* with climatic conditions modified due to elevation, is limited to the mountains of Ethiopia and the great African volcanos. Some parts of them rise to the snow line (Fig. 2.8).

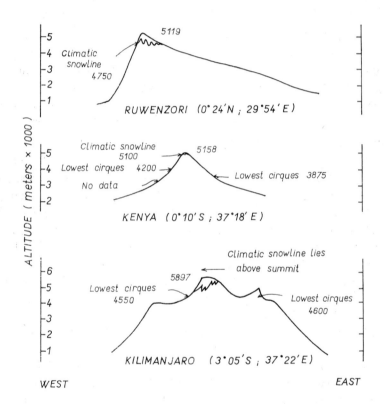

Fig. 2.8. Snowline of African tropical mountains.

Stamp's classification is in more ecological terms and is thus very convenient for hydrology; however, the lack of definition in terms of actual numbers makes it difficult to classify some basins near the margins of the climatic regions.

In terms of actual values, Köppen's classification is very useful. Köppen [10] recognizes:

Tropical rainfall climate with the mean temperature for all months above 18 °C. There are three subtypes defined as

— tropical forest climate with a minimum monthly rainfall of 60 mm
— monsoon climate
— periodically dry savanna climate. Here with an annual rainfall of

1000	1500	2000 mm,	the driest month should have
60	40	20 mm	or less of precipitation.

When there is a month with 0 precipitation, then the annual total should be 2500 mm or more.

Dry climate with two subtypes found in the tropics:

— steppe climate where the mean temperature of

25	20	15 °C	corresponds with a maximum annual rainfall of
700	600	500 mm.	

— desert climate where the mean annual temperature of

25	20	15 °C	corresponds with a maximum annual rainfall of
350	300	250 mm.	

Warm rainfall climate where the temperature of the coolest month ranges between −3 and 18 °C and a mean annual temperature of

5	10	15	20 °C	corresponds with maximum rainfall of
300	400	500	600 mm.	Three subtypes are found in relatively small localities

in the tropics:

— warm climate with dry winter
— warm climate with dry summer
— warm wet climate.

As can be seen from Fig. 2.9, a simplified classification is used to characterize each particular region:

A indicates tropical humid climates,
B indicates dry climates.

In general, *A* regions do not experience any frost and have an average temperature of over 18 °C. In the *B* regions evaporation exceeds precipitation and *B/A* marks the boundaries where the evaporation and precipitation are equal. For the zonation of types, *S* indicates occurrence of steppe, *r* is the rainy region with no more then two dry months and *w* indicates dry winter.

Thus the following subtypes are found in the tropics:

Ar   Tropical wet subtype,   *Ch* Warm climate with dry winter,
                                H Mountaineous climate.

Aw Tropical wet and dry subtype,
Bs Steppe and semiarid region,
Bw Desert and arid region.

A hydrologist concerned with the climate as a factor forming and influencing the hydrological cycle can summarise from this the following facts about the climatic tropical regions:

Fig. 2.9. Climatical regions of tropical Africa according to Köppen's classification.

Tropical humid climates form a belt less then 20° wide astride the equator which is mainly distinguished by absence of winter in the tropical lowlands. Frost is absent, except on the highlands and the plateau. The margins are determined either by diminishing annual rainfall or decreasing temperature. The belt merges with the dry climate in the western and central part of continent. Rainfall is abundant, coming mostly in short showers and thunderstorms. The two principal types within the tropical humid group are distinguished according to the seasonal distribution of the rainfall. While the tropical wet region has ample rainfall for ten or more months of the year, the tropical wet and dry region has rainfall for nine months or less and a dry season lasting more than two months. The wet climate region lies near the equator in the belt of permanent strong solar radiation and uniformly high temperatures. Annual temperature ranges between $10-25\,°C$ and the daily difference in temperature is frequently not more then $10\,°C$. This influences the diurnal evaporation/evapotranspiration to such an extent that, consequently, a diurnal fluctuation of the hydrographs can be traced on small streams. Due to the abundant cloudiness and humidity there is a very little cooling, though sufficient to produce fog and dew. Tab. 2.11 gives the basic data for Yaoundé lying in the wet climate region.

Tab. 2.11. Climatical data for Yaoundé (4°0′ N, 11°20′ E, 730 metres) in the tropical wet climate region

Jan.	Feb.	March	Apr.	May	June	July	Aug.	Sept.	Oct.	Nov.	Dec.	Annual
Mean daily range of air temperature (°C):												
10.0	10.0	10.0	10.5	8.9	8.3	7.8	8.4	8.3	8.9	9.4	9.4	9.4
Absolute maximum air temperature (°C):												
32.8	33.5	33.0	35.6	34.4	32.2	30.6	33.9	31.1	32.8	33.0	32.4	35.6
Absolute minimum air temperature (°C):												
16.0	15.9	16.9	14.5	17.0	16.3	16.0	16.2	16.5	15.5	16.8	17.5	14.5
Precipitation (mm):												
35	69	148	198	225	180	79	84	202	299	131	30	1680

Annual rainfall totals usually vary between 1750 and 2500 mm, and can be higher, where the air is more humid (Fig. 2.10). Rain-making disturbances are frequent, with abundant cumulus clouds producing heavy showers and thunderstorms. Most of the rain falls during the afternoon and early evening when solar heating has made the humid air increasingly unstable. Night rains are also frequent. Rainfall periods are rather short, continue for several days and are separated by periods of clear skies. Two subtypes are recognized in this climate. The basic subtype is characterised by the distribution of abundant rainfall throughout the year, the second is influenced by the monsoons with a more pronounced period, no longer than two months, and with precipitation of at least 60 mm per month. Some areas, usually considered to be parts of the second subtype, do not satisfy this condition, as can be seen from rainfall records in Freetown (Tab. 2.12).

Tab. 2.12. Monthly and annual rainfall (mm) at Freetown (8°20′ N and 13°05′ W)

Jan.	Feb.	March	Apr.	May	June	July	Aug.	Sept.	Oct.	Nov.	Dec.	Annual
10	8	29	101	225	290	510	900	930	725	130	36	3894

Tropical wet and dry climate has usually less annual precipitation, the rainfall is more seasonal, and there is a typical dry season of over two months. Dry and wet regions are located on the poleward side of the tropical climate and on the other side they border on dry climates. According to the orography and shape of the continent, the region is excentrically distributed around the equator: in the northern

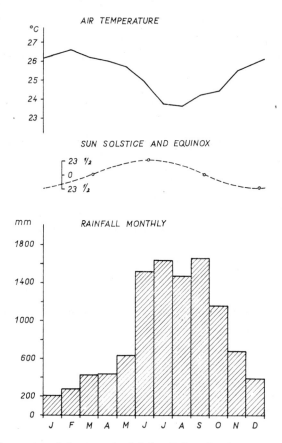

Fig. 2.10. Fluctuation of the annual rainfall at Debundja, Cameroon (4°N, 9°30′)

hemisphere not exceeding 15° north, in the southern hemisphere reaching almost to the tropic of Capricorn. The region is located between the ITCZ and its rain-bringing effects, and the subtropical anticyclones with their stable subsiding and diverging air masses on the poleward side. With the seasonal shifting of solar-energy belts and pressure and wind belts, the region comes under the influence of the moving ITCZ and its rain-bringing effects during high sun, while at the time of low sun it is under the influence of drier winds and subtropical anticyclones, both effects resulting in dry winter and wet summers.

In the tropical wet and dry region the annual range of temperature is greater — usually over 30 °C. The diurnal range is greatest in the dry season when the sky is clearer and humidity low, temperature regularity is still strong however, as can be seen from Tab. 2.13.

Tab. 2.13. Climatical data for Lusaka (15°25'S, 28°55'E, 1260 m altitude) in tropical wet and dry region

Jan.	Feb.	March	Apr.	May	June	July	Aug.	Sept.	Oct.	Nov.	Dec.	Annual
Mean diurnal range of air temperature (°C):												
8.4	8.9	8.9	11.1	12.8	12.8	13.4	13.3	13.9	13.3	11.1	9.5	11.1
Absolute maximum air temperature (°C):												
21.1	30.6	30.0	30.6	29.4	28.3	28.3	30.6	35.0	37.8	36.7	33.9	37.8
Absolute minimum air temperature (°C):												
14.4	13.3	12.8	10.0	8.3	3.9	4.4	6.1	7.8	12.2	12.8	13.9	3.9
Precipitation (mm):												
231	190	142	18	3	2	2	0	2	10	91	150	835

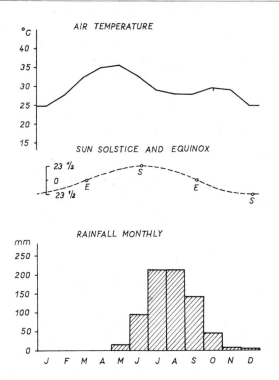

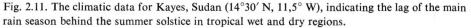

Fig. 2.11. The climatic data for Kayes, Sudan (14°30' N, 11,5° W), indicating the lag of the main rain season behind the summer solstice in tropical wet and dry regions.

The variability of diurnal temperature is also strong. Often the hottest period precedes the time of highest sun (Fig. 2.11). Thus March, April and May are usually warmer than June and July, which are the rainy months north of the equator, and September, October and November are usually warmer than December and January in the southern hemisphere. Annual rainfall for the region is usually given as varying between 1000 and 1500 mm. As can be seen from Tab. 2.13, lower annual rainfall is common. Generally, the annual rainfall is smaller than in tropical wet regions because of the transitional type of the climate between wet and dry. The rain comes in the form of convective showers as in the wet regions. A return of ITCZ toward the equator brings the anticyclonic drought in winter. Thus, the farther the region is from the equator, the longer is the dry season.

In the southern parts of the continent there are extensive upland areas, which have somewhat lower temperatures resulting from the greater elevation. Monthly temperatures may thus occasionally be below what is considered the minimum for tropical climates, nevertheless, the areas are still classified as tropics. For instance, Sesheke on the Middle Zambezi has temperatures below freezing point ten days a year. As will be seen later, the difference between lowlands and uplands is very important since the total "loss" in the uplands is considerably lower.

Highland climates cannot be characterized uniformly. The type of climate of a particular mountain or mountainous region depends on the altitude, exposure to sun and wind and the latitude. Even isolated small valleys may have a considerable variability in climate on the leeward and windward slopes. Generally it can be said that the highland climates are low-temperature variants of low elevation in similar latitudes. An accepted decrease of the mean temperature is $0.65-0.95\,°C/100$ m. An example of the highland regime is given in Tab. 2.14:

Tab. 2.14. Climatical data for Addis Ababa (9°29′ N, 8°43′ E, 2440 m) in tropical highland region

Jan.	Feb.	March	Apr.	May	June	July	Aug.	Sept.	Oct.	Nov.	Dec.	Annual
Mean diurnal range of air temperature (°C):												
17.8	16.1	15.6	15.0	15.0	13.9	10.6	10.6	12.8	16.7	16.7	17.8	15.0
Absolute maximum air temperature (°C):												
27.8	30.0	28.9	31.1	32.8	34.4	31.1	28.9	27.2	32.8	27.2	27.8	34.4
Absolute minimum air temperature (°C):												
1.7	2.2	3.3	4.4	3.9	6.7	7.2	6.1	3.3	2.2	0.6	0.0	0.0
Precipitation (mm):							*u*	*ň*				
14	37	70	85	90	134	285	295	195	21	13	6	1246

The influence of sunlight intensity increases in the highlands, because the dust and clouds are concentrated at lower elevations of the tropics. The thin, dry air of the mountains and high plateaux permits not only the entry of strong solar radiation by day, but also rapid earth radiation at night. This results in many days with nocturnal freezing and thawing by day in the tropical mountains. The great difference between daily maxima and minima is also in contrast to the very small differences between the maxima and minima of the warmest and coolest month. Generally, tropical mountains have heavier precipitation than the surrounding lowlands.

Dry climates are found in the regions where a water deficiency prevails. Since the evaporation/evapotranspiration rate depends on the temperature, the deficiency is higher in warmer climates.

The semiarid type of climate is in a transitional belt surrounded by desert and separating it from the humid climates. In tropical Africa, it is found anywhere between the equator and the tropics. According to Trewartha, the climate passes through the latitudinal zones and thus is not defined in terms of temperature, unless the rainfall is taken into account. The rainfall in semideserts is slightly higher then in the deserts. Total annual rainfall is sometimes limited to 250–500 mm, according to some sources, to 375–600 mm. The tropical steppes have a short period of heavy rain at the time of high sun. The rainfall distribution is more or less similar to the wet and dry climate except that the dry season is longer and total precipitation less. Belts of steppe on the poleward side of the deserts are usually in fairly close proximity to the subtropical dry summer type and have nearly all their rainfall in the cool season, with middle-latitude cyclones which tend to travel equatorward more in winter than in summer. Tab. 2.15. gives data typical for the semiarid region.

Tab. 2.15. Climatical data for Windhoek (22°37' S, 17°8' E, 1228 m) in the tropical semidesert region

Jan.	Feb.	March	Apr.	May	June	July	Aug.	Sept.	Oct.	Nov.	Dec.	Annual
Mean diurnal range of air temperature (°C):												
12.2	12.2	11.7	12.2	13.3	13.3	13.9	14.5	13.3	13.9	13.9	13.3	12.8
Absolute maximum air temperature (°C):												
36.1	34.4	34.4	30.6	31.7	26.1	25.0	29.4	32.8	33.9	35.6	36.1	36.1
Absolute minimum air temperature (°C):												
9.4	6.7	3.9	2.2	1.7	2.8	2.8	3.9	0.6	1.7	0.6	3.3	0.6
Precipitation (mm):												
77	74	83	40	7	1	1	1	2	12	22	48	368

In the arid climate regions the summer is very hot and winter is cool. The extreme conditions are caused by the leeward and interior location of most dry regions and by the prevailing clear skies and dry atmosphere. An abundance of solar energy reaches the earth by day and is rapidly lost during the night. Rainfall is always insufficient and very variable. There are more years with rainfall below average than above. There is a higher average during a very humid year, however, because the precipitation takes the form of storms, the water flows away as floods, and little excess remains for vegetation. A humidity of 12%–13% is usual for the midday hours. Potential evaporation is therefore high and the wind contributes to its further increase. Dry regions are often windy, influenced by subtropical anticyclones and dry trades. The tropical deserts receive less then 250 mm and sometime even areas with 375 mm are considered to be deserts. In some locations there is no rain for several years. Showers fall at any time of the year but very irregularly. Tab. 2.16 gives some basic data for a desert locality.

Tab. 2.16. Climatical data for Tessalit (20°12′ N, 1°00′ E, 520 m) in the tropical desert region

Jan.	Feb.	March	Apr.	May	June	July	Aug.	Sept.	Oct.	Nov.	Dec.	Annual
Mean diurnal range of air temperature (°C):												
14.5	15.6	15.5	15.0	15.0	14.5	14.5	13.8	14.8	13.3	13.9	13.9	14.4
Absolute maximum air temperature (°C):												
32.8	37.2	39.6	40.0	44.4	46.3	45.6	44.4	44.4	41.7	38.3	35.0	46.3
Absolute minimum air temperature (°C):												
4.2	4.4	8.9	11.3	15.8	21.0	20.0	18.9	17.8	15.0	11.1	3.3	3.3
Precipitation (mm):												
0.8	0.8	2	1	4	15	16	64	38	4	0.4	0.5	64

The low-altitude desert areas have average hot-month temperatures between 30°–35° C and cold-month between 10°–15° C. Altitude modifies this range. In some locations, temperatures over 70 °C have been recorded. The diurnal range may reach 35 °C and except for the marginal parts, the range is similar from day to day.

A modification of the dry climate, the so-called *Marina dry climate,* is found at some places along the west coast, south of the equator, and at the Somali coast. The climate is under the influence of the cool ocean current and is characterized by markedly lower summer temperature, a reduced range of annual temperatures and a reduced range of daily temperatures. Rainfall is even lower than in the nearby desert, due to the subtropical anticyclone and the stabilising effects of the cool coastal water. Fog and low cloudiness contribute toward lowering the potential evaporation rate.

2.5 LIST OF LITERATURE

[1] Aune, B., 1970. Tables for computing potential evaporation from the Penman equation; Zambia Meteorological Department, 37 p.

[2] Bailey, M., 1969. On extreme daily rainfall in Zambia. Department of Meteorology, 10 p.

[3] Bailey, M., 1970. Rainfall probability in Zambia. Department of Meteorology, Lusaka, 12 P.

[4] Balek, J., 1971. Daily fluctuation of potential evaporation. National Council for Scientific Research Report TR 16, Zambia, 49 p.

[5] Balek, J., 1972. Use of representative and experimental basins for the assessment of hydrological data of African tropical basins. Symposium on the design of water resources projects with inadequate data, Madrid, 17 p.

[6] Barry, R. G., 1969. Precipitation. Water, Earth and Man. Methuen and Comp., London.

[7] Fosberg, F. R., Ganier, B. J., Küchler, A. W., 1961. Delimitation of Humid Tropics. Geog. Rev., 51, pp. 333—347.

[8] Garbell, M. A., 1947. Tropical and equatorial meteorology. Pitman, New York, 237 p.

[9] Herbetson, A. J., 1905. The major natural regions. *Geogr. Journal* XXV, pp. 300—312.

[10] Köppen, W., Geiger, R., 1954. Klima der Erde. J. Perthe, Darmstadt.

[11] Lebedev, A. N., 1967. Klimaty Afriky (Climates of Africa). Hydromet. Publishing Comp., Leningrad, 488 p.

[12] Lee, D. H. K. Climate and economic development in the tropics. Harper and Bros., New York, 182 p.

[13] Pollock, N. C., 1968. Africa. University of London Press.

[14] Pritchard, J. M., 1969. Africa. Longmans, 268 p.

[15] Raman, P. K., 1971. Rainfall climate of the project area. Hydromet. Survey of Lakes Victoria, Kyoga, Albert, Entebbe, 51 p.

[16] Riehl, H., 1959. Tropical meteorology. McGraw Hill, New York.

[17] Stamp, L. D., 1965. Africa: A study in tropical development. J. Wiley and Sons, New York.

[18] Strahler, A. N., 1975. Physical geography. J. Wiley and Sons, New York, latest edition.

[19] Trewartha, G. T., 1943. An introduction to climate. McGraw Hill, New York, 402 p.

[20] Voeychkov, A., 1883. Flüsse und Landseen als Produkte der Klimate. From ,,Die Klimate des Erdballes", St. Petersburg.

3. HYDROLOGICAL CYCLE AND WATER BALANCE OF A TROPICAL BASIN

3.1 GENERAL CHARACTERISTICS

The role of the hydrological cycle in the tropics is very similar to its role in moderate regions, although several factors affecting the water movement from the atmosphere through and over the earth into the rivers, are of special importance and resulting effects deviate from those experienced outside the tropics. The function of the vegetation primarily affects the tropical cycle in addition to the sun. Both natural vegetation and introduced plants play a significant role and, with a view to the rapidly advancing agriculture in Africa, land-use problems require many answers from hydrologists. The promotion of industrial development and civil engineering also requires more accurate hydrological data based on detailed knowledge of the water balance in a given region. Natural conditions, different from those in Europe and the USA, do not allow even basic problems within the tropics to be solved by direct application of formulae and routines developed for moderate regions. For instance, the special function of interception, different time and space distribution of rainfall, the pronounced influence of swamps, soil problems etc., require comprehensive research into problems already solved elsewhere. Many results have been already obtained, but there are still many problems to be solved.

3.2 VEGETATION AND ECOLOGY

An ecological approach to the management of natural resources, and particulary water resources, has not yet been widely adopted in tropical Africa, partly due to lack of basic knowledge of many fundamental aspects of the structure, functioning and dynamics of the tropical hydroecosystems. As for ecology, several attempts have been made to classify the ecosystems with regard to the function of the water involved. Oversimplified, there are two basic hydroecosystems in tropical Africa, namely the humid tropics and the arid tropics. In a more complete classification, one can differentiate between systems with water deficiency under dry conditions, systems with water excess under wet conditions and systems with combined effect under wet and dry conditions. According to Kimble [14], Africa can be divided into the following hydroecological regions:
1. Perennially well-watered regions,
   a) rain forests,
   b) perennial swamps,
   c) great lakes.

2. Seasonally well-watered regions,
    a) savanna,
    b) dry forests,
    c) seasonal swamps.

3. Perenially ill-watered regions,
    a) desert,
    b) semidesert.

In addition to climate, this classification is dominated by vegetation.

Rainforest is an area where the annual rainfall is in excess of the water needs of plants, even if the water is not more than the forest can use. Under some conditions, the runoff potential may be greater than infiltration into the soil permits, however, in a mature forest, there is a continuous downward movement of water and continuous transport of water through the vegetation. After artifical cleaning or natural fires, when the soil becomes exposed to erosion and solar radiation, the balance of the forest is disturbed. The water table is lowered, depending on the rate of infiltration, or raised, depending on the smaller amount of water emitted in transpiration. An upset ratio between the components not only directly affects the area but also the valleys. Some of the most significant rivers of tropical Africa are fed from the rain-forest area: they are the Middle Niger, Tana, Rujiji and the most important subbasins of the Congo.

Perennial swamps are frequently found in the flat areas of tropical Africa. As a special chapter will deal with swamps, only their function as a runoff regulating system will be discussed here. During the dry months, they are able to release more water than receive, even if not in the most economic way. Frequently more water is lost by evaporation than by the runoff. Remarkably, swamps are not related to a single climatic zone. They are rather widely distributed in the tropics, tending to be more numerous near the arid tropical margins. The swamp vegetation plays a specially important role in hydrology and morphology and influences the hydro-logical cycle to a great extent.

Lakes occupy much less of the tropics than the swamps and several of them are saline or brackish, which reduces their utility for agriculture. They include the lakes Rukwa, Manyara, Eyasi, Balangida, Natron and Magadi. The outflow varies from lake to lake, some, such as the Rudolph, are without any outflow, sometimes the outflow is small to negligible (Tanganyika) and sometimes they are drained by mighty streams such as Victoria or Nyasa. The hydrological cycle is specially complicated in the marginal areas between lake and land where extensive islands of aquatic vegeta-tion are found. Does this part belong to the lakes or to the land? This question must always be answered before the water balance of a lake or its tributary is calculated. Sometimes, as at the outflow of the Shire river from Nyasa lake, the vegetation blockage and silt deposits influence the long-term vegetation and regime of the lake. After an occasional breakthrough, the lake tends to decrease until a new barrier of vegetation and deposits is formed.

Forest and savanna cover approximately one half of the African tropics. This means that half of the tropics suffer from alternating surpluses and shortages of water. There are several definitions of savanna. According to Beard [9] savanna is the natural vegetation on the highly mature soils of senile landforms which are subject to unfavourable draining conditions and which have intermittent perched water tables with alternating periods of waterlogging (with stagnant water) and desiccation. Walter [24] calls the natural savanna an ecosystem in which the grassed and woody plants compete for water and the grasses are the dominant partner. In many types of savanna a differentiation is made between grass savanna, shrub savanna, tree savanna and savanna woodland. Water plays a dominant role in any classification since the combination of grass, woodland and bush is adaptable to alternating conditions of water surplus and deficiency. Savanna signalises the amount of water stored by the degree of greenness.

Seasonal swamps differ from perennial swamps in the duration of the outflow, their size and the vegetation. The presence of a water level is also an important characteristic, since some of the swamps last only for weeks, others may become temporarily perennial and others may be ephemeral. Some of the intermittent swamps have developed on slopes where a large population of rooted plants has slowed down the surface movement and provided favourable conditions for swamp formation.

Semideserts experience rainfall only occasionally and usually in a torrential form. The rain water can flow rapidly underground and some authors estimate that about 50% infiltrates into the semideserts, where the soil structure, may or may not protect it from evaporation. Some water is also deposited in the river beds.

The desert has water of an exotic origin, unknown in most cases. Some may travel by underground courses, some may be of an older date. Actually, very few deserts exist without some water stored at some depth. In some desert, such as on the coast between the Orange and Cunene rivers, fog is a dominant part of the hydrological cycle. Occasionally the fog contributes significantly to the hydrological cycle of the deserts, supporting some short-lived plants. Similarly, in the Sahara, nightly falls of dew during the cooler months of the year are sufficient for the growth of some plants.

Worthington [28] classifies the so-called aquatic ecosystems more or less on a regional and zoological basis. In this respect he talks about Nilotic, Congoan, Zambezian, Victorian, Tanganyikan and Nyasan system. He supposes that in the Miocene these systems were in contact, but now their fauna is distinctive, because of the tectonic movements, the formation of the Eastern and Western Rift Valleys, and the depressions of Lakes Victoria and Kyoga. Volcanic activity also diverted some streams and the pluvial and interpluvial periods contributed to the formation of distinctive faunas found nowhere else in the world. The flora, however, has not become so distinctive, probably because of the ability of seeds to be transported by wind, water and man.

The African vegetation does not follow the pattern of the isohyetal map. For instance, the savanna in some parts of southern Nigeria and northern Zambia is adapted to a 1500 mm rainfall and up to eight months of rainfall in the regions with a high level of groundwater, but also to areas with 600 mm of rainfall and a low groundwater table. Thus, it is not a simple problem to classify the vegetational belts and several systems have been developed. Fig. 3.1 presents a modification of a rather simple system which may be found satisfactory for hydrological purposes:

Mountain forest and grassland (the more extensive) are located in the African highlands above 1500 meters. These belts are less significant than the others.

Fig. 3.1. Vegetational belts of tropical Africa, after Shantz. 1 — Equatorial rain-forest, 2 — Woodland and savanna, 3 — Wooded steppe, 4 — Semidesert steppe, 5 — Desert, 6 — High plateau grass and forest, 7 — Modified equatorial rain-forest.

52

Equatorial rainforest occurs discontinuously in equatorial latitudes with abundant precipitation and constantly high temperature. Evergreen forests mainly cover the Congo basin and have several distinctive layers.

Savanna, forest savanna and woodland savanna form a vast belt of vegetation in the zone of seasonal rainfall that almost surrounds the rainforest and reflects the various ratios of the length of the dry and wet seasons.

Dry forest is found south of the equator on the plateaux of East and Southeast Africa. Wide areas of the Miombo forest lie in this belt. Owing to the increasing influence of man, the dry forest is rather limited in West Africa.

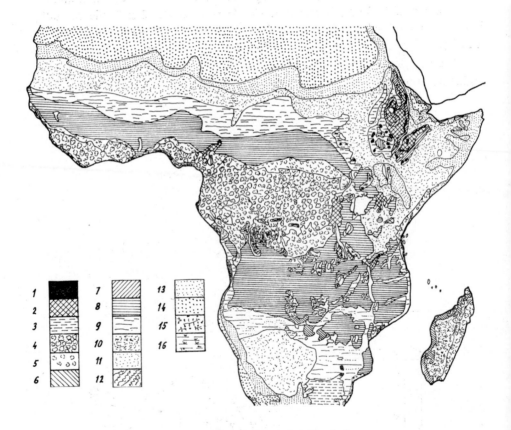

Fig. 3.2. Vegetational belts of tropical Africa, after Aubréville et al.
Main vegetation belts: 1 — Montane evergreen forest, 2 — Montane communities (undifferentiated and mountain grassland, 3 — Temperate and subtropical grassland, 4 — Moist forest at low and medium altitudes, 5 — Forest-savanna mosaic, 6 — Dry deciduous forest (and savanna), 7 — Thickets, 8 — Woodlands and savannas, relatively moist types, 9 — Woodlands and savannas, relatively dry types 10 — Grass savanna and grass steppe, 11 — Wooded steppe, 12 — Other grass steppe, 13 — Subdesert steppe, 14 — Desert, 15 — Mangrove, 16 — Swamps.

Acacia savanna and thorn forest of semidesert steppes is found in the areas with seven to eight months of dry season and a rainfall below 500 mm. Conditions are too dry for the majority of perennial grasses and acacia species have adapted themselves by thin leaves, thick barks and deep roots. At the desert, marginal parts of long and narrow vegetation, formed by *desert shrub* and *grass,* graduate into pure *desert.* Ephemeral grasses and wide scattered patches of scrub are the only types of vegetation surviving in the desert.

This classification, based on Shantz's description, has been criticized as failing to emphasise the essential unity and continuity of the vegetation of tropical Africa. A more detailed classification of vegetation has been published with Unesco assistance [1] and is presented in Fig. 3.2. A comparison of the climatological and vegetational belts shows that the latter follow the climatic changes only to a certain extent. Actually, any small-scale vegetational map of an African tropical region indicates a very high variability of originally introduced vegetational cover, even within small areas, and the task of finding a fully representative catchment for a given vegetational belt may be found to be difficult if not impossible.

Hydrologically, a very convenient ecological classification was published by Whyte (Tab. 3.1) [26]. He includes natural characteristics as well as the prevailing types of introduced plants and the influence of man.

3.3 THE ROLE OF TROPICAL FOREST IN THE HYDROLOGICAL CYCLE

The relationship between vegetation and hydrology has been studied for a long time and some significant results have been achieved. The role of the African forest has been under intensive study since the forest covers approximately 25% of Africa.

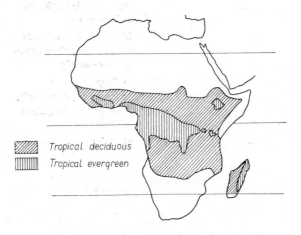

Fig. 3.3. Forest belts of tropical Africa, after Matějka.

Tab. 3.1. Ecoclimatic gradient in tropical Africa; after Whyte

Zone	Rainfall range (mm)	Dry season month	Vegetation	Forest formation
Saharan	—	12	Desert	—
Subsaharan	10—250	12	Subdesert steppe	—
Sahelian	250—600	9	Wooded steppe	Shrub, thorny savanna
Sudanian	600—1250	6—8	Woodland, dry savanna	Open woodland (tree sav.)
Guinean	above 1250	3—6	Woodland	Seasonal Guinean forest
Guinean Equatorial	above 1800	short	Moist forest at low and medium alt.	Closed rainforest
Eastern Equatorial Desert/Subdesert	250	12	Desert, subdesert steppe	—
Eastern Equatorial Savanna	250—600	9	Wooded steppe	*Acacia Commiphora* woodland
Eastern Coastal	600—1250	6—8	Coastal forest savanna mosaic	
Lake Victoria	1250	—	Forest savanna mosaic	
East and Central	1000	5—6	a) Woodlands of various types *Brachystegia* and *Julbernardia* b) Savanna woodlands with *Acacia, Combretum* and *Terminalia* c) *Cryptosepalum pseudotaxus, Marquesia, Guibourtia coleosperma* on Kalahari sands d) *Colophosperum mopane*	Miombo and Mopane woodlands
Eastern African Highlands	1000		Montane grassland and forest	

Land use	Crops	Countries
Nomadic grazing	—	Mauretania, Mali, Niger, Chad, Sudan
Nomadic grazing	—	Mauretania, Mali, Niger, Chad, Sudan
Seminom. grazing, semiarid cult.	Sorghum, millet	Senegal, Mauritania, Mali, U. Vola, Niger, Chad, Sudan
Seminom. grazing and arable cult.	Sorghum, millet, groundnuts, yams, maize	Senegal, Mali, U. Volta, Niger, Northern Nigeria, Chad, Sudan
Arable cult., forest utilization.	Sorghum, millet, groundnuts, cassava, yams, maize, irr. rice	Senegal, Gambia, Guinea Bissau, Guinea, Mali, Ivory coast, Upper Volta, Ghana, Togo, Dahomey, Nigeria, Cameroon, Chad, Central African Rep., Sudan, Congo, Zaire.
Arable cult., tree crops, forest util.	Rice, bananas yams, taro, maize, oil palm, cocoa, rubber, coffee	Sierra Leone, Liberia, Ivory Coast, Ghana, Nigeria, Cameroon, Gabon, Congo, Zaire
Nom. grazing	—	Somalia, Ethiopia, N. Kenya
Nom. grazing and cultivation subsidiary to herding	Millet, sorghum	Somalia, Ethiopia, Sudan, Kenya, Tanzania
Arable cultivation	Rice, cassava, coconut	Somalia, Kenya, Tanzania, Mozambique
Grazing and arable cultivation. Tree crops, mainly coffee.	Plantains, bananas, millet, sorghum, coffee, cotton	Kenya, Uganda, Tanzania
Arable cultivation, tobacco planting, Europian farming	Millet, sorghum, cassava, maize, tobacco, pastures	Tanzania, Zaire, Angola, Zambia, S. Rhodesia, Malawi, Mozambique
Grazing, arable cult.	millet, cassava, maize, sorghum, sown pastures	
Grazing and arable cult.	Millet, cassava, maize	
Grazing and ranching. Arable cult.	Maize, sorghum, millet	
Grazing and arable cult. Ranching, mixed farming.	Maize, sorghum, plant., bananas, coffee, tea	Ethiopia, Kenya, Rwanda, Burundi, Tanzania, Congo, Malawi, S. Rhodesia

The forest is not distributed uniformly over the whole continent. The forest cover is considerably larger in the tropics. Actually, as can be seen from Tab. 3.1, the term "forest formation" is more appropriate, since there is a great variety of forest types. The map in Fig. 3.3 provides an approximate estimation. There are many types of forest in West Africa, where the three basic types are found: tropical rainforest, mixed deciduous forest and savanna forest. Gabon is entirely covered by forest and tropical forest covers significant parts of Cameroon, Nigeria and southern Ghana, the Southern Ivory coast, Liberia and Zaire. High species of *Entandrophragma*, *Khaya*, *Anopyxis*, *Erythropleum*, *Lophira*, *Piptadenia*, *Tarrietia*, *Aucoumea*, *Decroydes*, *Terminalia*, *Pycnanthus* are dominant there.

Mixed deciduous forest forms a belt north of the tropical rainforest. A considerable percentage of deciduous species is found here besides evergreens.

The savanna forest is distributed in Ghana, the Ivory Coast, Upper Volta and Guinea. Under man's activity this forest has deteriorated in many parts of Africa: *Daniella Oliviery*, *Pterocarpus*, *Erinaceus*, *Isoberlinia*, *Lophira Lanceolota*, *Parkia* and various palms are still plentiful.

In Central Africa the tropical rainforest, covering the Congo basin, prevails. *Macrologium dewevrei*, *Julbernadia seretii*, *Cynometra alexandri*, *Celtis soyauxii*, *Brachystegia laurentii* and other species are dominant. At the margins of the area where two to four months of rainless period occur, the wet, semideciduous forest consists of *Entandrophragma*, *Fagara*, *Guarea*, *Khaya*, *Afzelia*, *Milleti* and some others. Outside the margins, dry deciduous forests are found, influenced by the occurrence of dry winds.

In Southern Africa, wet evergreens of the subtropics are found: *Olea*, *Ocotea*, *Curtisia*, from coniferous *Podocarpus*, *Widringtonia*. Numerous types of pines were introduced in East Africa. Owing to the low precipitation and of mountain shadowing, some semi-dry areas are covered by succulents as *Sansevieria* and *Euphorbia*. The dense forest covers the areas between 1600 and 2700 meters. Lower parts of this belt are covered by *Entandrophragma*, *Ocotea*, *Lebrunia* and *Fagara*. Above the *Podacarpus usambarensis*, conifers Pine and Juniper are found. A great part of the wet and dry tropics is covered by the open savanna forest "Miombo" with *Isoberlinia*, *Acacia*, *Brachystegie Uapaca* and baobabs. Most typical for great parts of Angola, Zambia, Tanzania, Malawi, Mozambique and Botswana are open forests with *Brachystegia*, *Isoberlinia* and *Burkey*; various pines and eucalyptus have been introduced. The influence of the African forest on the hydrological regime has been studied for a long time. Wicht [27] drew up a set of conclusions based on observations in South Africa:

a) Plantations of exotic trees require approximately the same amount of water as the indigenous forest.

b) Forest will use more water than grass.

c) The consumption of water by forest depends primarily on the amount of water available in the soil.

d) Swamps will dry out if trees are planted in them.

e) The removal of vegetation will increase the discharge.

A basic problem still to be solved in the relationship between the forest and the hydrological cycle, has been formulated by Barnard when analysing the Congo basin: Is the high rainfall above Congo a result of the forest or is the forest a consequence of the high rainfall? There is no definite answer and observations from the small experimental areas cannot supply adequate information. The catchment experiments, however, are very valuable for supplying answers to numerous equally important questions. Pereira et al. [21] established catchments on the volcanic mountain ranges which border the Rift Valleys in Kenya. When exploring the role of the vegetational cover they found that even if the deep permeable soils are covered by heavy forest, the seasonal streamflow fluctuation is 1 : 20.

On the same catchments, experiments with land changes indicated that the plantation of pine in the bamboo forest led to a substantial increase in the water yield. Cropping with peas, beans, potatoes and squashes reduced the water consumption by 350 mm as compared with bamboo cover. While the ratio of E_t/E_0 (actual evapotranspiration to potential evaporation) for bamboo was 0.9, for vegetables and pine seedlings it was only 0.51, maize and pine 0.72 and for patula pine samplings it was 0.78. As reported later [20], tea plantation increased the runoff on the area cleared of forest. The experiments have proved that bamboo and or tall montane forest are an ideal protection of the surface against surface runoff. For coffee plantations the ratio E_t/E_0 was found to vary between $0.5-0.8$, while at the same time for a tropical forest it was 0.86.

The year by year fluctuation of both values, measured elsewhere, indicates that additional parameters, such as groundwater fluctuation and seasonal distribution of rainfall, may play some significant role. As proved by Dagg and Pratt [10] on the catchments Sambret and Lagan in Keyna, forest cover makes the catchments almost immune against streamflow, regardless of the effects of rainfall intensity and soil moisture. They observed less then 10% of the total outflow in the form of overall streamflow. Hellwig [11] in a study on evaporation from sand, puts 68% of the total water loss from sandy river beds in South West Africa to the account of the phreatophytes *Acacia albedo,* and *Tamarix austro-africana.* Various observation from arid regions confirm that a substantial increase in water yield occurs after replacement of brush vegetation by grass. The deep-rooted aspen can evaporate twice as much as barren soil. Hellwig estimates that about $190-500$ mm of water can be saved in arid regions by the removal of phreatophytes.

It can be concluded that trees can adapt themselves under tropical conditions to a very variable amount of available water. A widely accepted assumption that trees can live exclusively on the water available in the soil was changed when the results from the Luano experimental catchments became available [7]. Observations taken there indicated that the groundwater level through the capillary zone provides additional water for the trees during the dry season. The total amount of water transpired from a tropical catchment may exceed the precipitation for that year if the previous year was rich in precipitation and the following one average or below normal.

The influence of felling tress on the hydrological regime gives only limited information if executed on a small scale within an otherwise undisturbed environment. Effects obtained on a large scale may supply very different results and produce another type of influence on the hydrological cycle than can be seen in a small-scale experiment. As stated by Whyte [26], the disappearance of the woody plants in the tropics and reduction in their average height worsens the microclimates, because the action of wind and plants and soils becomes more marked, temperature variation becomes greater on and below the soil surface, infiltration is affected and evaporation is more intensive. This explains why trees and shrubs are no longer reproduced. This is even more important with regard to the effectiveness of desert fringe types of vegetation acting as a protective barrier between the desert and the subhumid and humid regions.

3.4 THE ROLE OF GRASSLAND AND SAVANNA IN THE HYDROLOGICAL CYCLE

A transition from pure grassland to savanna and tropical dry forest, and its dependence on the amount of rainfall per annum, is shown in Walter's diagram developed for deep fine sandy soils (Fig. 3.4). Walter [24] assumes that with a 100 mm of rainfall per annum, only a 1 m layer becomes moist and only the shallow roots of grass may grow. With up to 300 mm, no water remains in the layers and only when 300 mm or more is accumulated in a year, are conditions favourable for shrub and tree growth and tree savanna originates. There is another limit, say 500 – 600 mm, at which the trees form a closed crown canopy and grass becomes a tolerated species. Then savanna changes into tropical forest, usually of the mopane type.

As correctly stated by Walter, such a state of equilibrium is found only rarely and under natural conditions. Besides rainfall, the soil moisture and groundwater conditions play an equally important role. As has been proved experimentally, the soil moisture itself may not be sufficient to support a mature forest and some groundwater table within the reach of the roots is required by the forest. On the other hand, many localities of savanna of anthropogenic origin are found in the tropics because of the frequent burning of vegetation and attempts to cultivate the soil. On the contrary, man-made desert is frequently formed in localities of natural savanna due to overgrazing. Under natural conditions, both precipitation and the whole water economy are the decisive factors influencing the formation and disappearance of grassland and savanna.

The trees check the transpiration through the leaves and when the closing of the stomata is not effective enough, the leaves are shed. Even then, some species of shrubs without leaves may transpire up to 8% of their fresh weight per day. This means that the woody plants always need some water to survive the prolonged dry season. On the other hand, during the dry season grass needs only a very small amount of water.

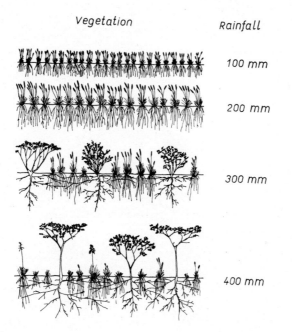

Fig. 3.4. Scheme of the transition from pure grassland to savanna and tropical dry forest with increasing rainfall, after Walter.

The conclusion drawn on the decrease in vegetation cover with decreasing rainfall can be also extended to the areas of the leaves according to Walter's rule: "The transpiration surface of the grasses decreases in proportion to the rainfall so that the water supply of a unit transpiring surface is of the same order under extremely arid or more humid conditions" [24]. Thus a linear function between the productivity of grassland in relation to dry weight and the average yearly rainfall has been developed for Southwest Africa (Fig. 3.5). Unfortunately, analogous graphs for higher rainfall values and for wood and forest dry-weight productivity are not available.

Thus, in arid regions a differentiation is made between

a) diffusive types

b) restricted types

of vegetation. Whereas the first are found in wide areas of arid regions, the second are restricted to depressions, wadis, oases etc. where more rainfall is concentrated than is typical for arid regions and exceeds the amount typical for semideserts.

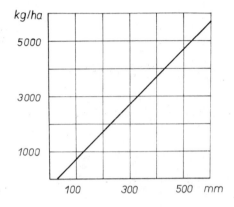

Fig. 3.5. The productivity of grassland in relation to the yearly rainfall.

3.5 THE ROLE OF ROOT SYSTEMS IN THE HYDROLOGICAL CYCLE

The roots of tropical species are well adapted to the conditions of the local hydrological cycle and may frequently serve as indicators of the subsurface situation in the fluctuation and exchange of water stored below the soil surface.

Tropical grasses with a rather high transpiration rate have a very dense root system. No less than 80,000 metres of total root length penetrating no more than 1.5 meter deep have been estimated for some single grass plants.

It is much more difficult to trace the tree roots because of the amount of labour needed to clear the root systems. A new view of the role of groundwater in relation to tree evapotranspiration in the Miombo forest has been based on the observations and measurements of Maxwell [19]. An analysis of the root system down to a depth of seven metres clearly indicated the relationship between the root density, groundwater level and soil moisture in saturated and unsaturated zones. In Fig. 3.6, the pictures of the root systems of dominant species indicate that most of the roots are situated in the upper soil layers, however, the density of the distribution varies from species to species. A secondary system reaches the saturated zone and takes the water through this zone from the groundwater storage, during the period of water deficit in the top layers. This has been proved by soil-moisture measurements taken in ten-day periods to the lowest groundwater level for several years. Fig. 3.7 gives a few measurements for some dominant soils in the studied area. Fig. 3.7a presents the measurements in poorly-drained, slowly permeable, very strongly to strongly

Fig. 3.6. Root systems of some dominant species of Central African Plateau (a—d).
a) *Brachystegia Longifolia*, b) *Julbernadia*,

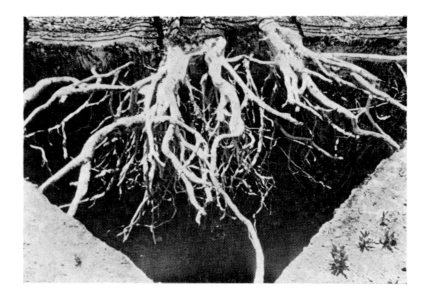

c) *Marquesia,*

d) *Brachystegia Microphylla.*

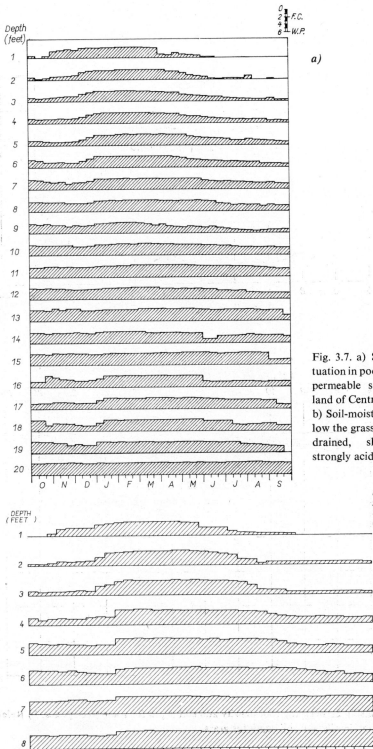

Fig. 3.7. a) Soil-moisture fluctuation in poorly drained slowly permeable soils under woodland of Central African Plateau. b) Soil-moisture fluctuation below the grass cover. Soil poorly drained, slowly permeable, strongly acid.

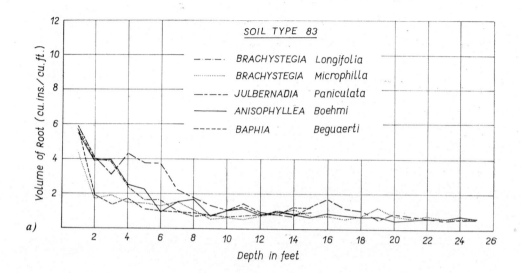

a)

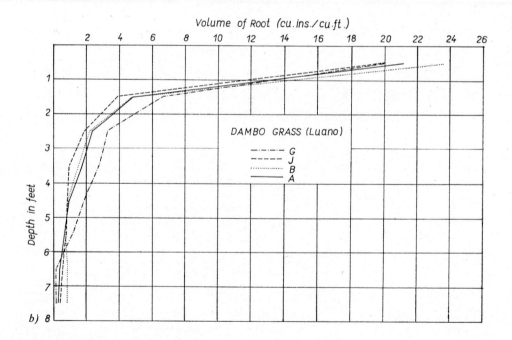

b)

Fig. 3.8. a) Relationship between the root density and soil depth for some tree species. b) Relationship between root density and soil depth for grass.

acid soil, consisting of sand, loamy sand and sandy loam. The depletion of the saturated and nonsaturated zones is clearly visible in the illustration. Another record, from a soil below grass cover which is poorly drained, slowly permeable, strongly acid and with a groundwater table less then 1.5 meters below the surface, is given for comparison. (Fig. 3.7b). Here the shallow grass roots tap only the upper moisture layers. The distribution of the density of the tree roots and grass roots as plotted in Fig. 3.8 is remarkable. A secondary increase in the saturated zone of the dominant tree species can be seen in addition to the high density of roots in the upper layers. On the other hand, the root density of the grasses rapidly decreases with depth.

Similar results were achieved by Lawson et al. [17] in Ghana. The root systems of the dominant species *Terminalia, Burkea, Detarium Pilostigma* and *Coclospermum*, of Guinean savanna, were studied as shown in Fig. 3.9. Again, a relationship between

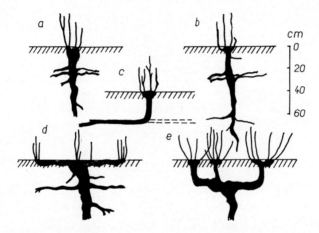

Fig. 3.9. Roots of some species in the Guinean savanna. a) *Terminalia Avicenniodies*, b) *Burkea Africana*, c) *Detarium Microcarpum*, d) *Piliostigma Thonningii*, e) *Cochlospermum Planchonii*.

the roots and the water regime can be traced, although comparable soil-moisture measurements are not available.

3.6 INTERCEPTION

The vegetation also significantly affects interception in the tropics. This phenomenon plays the role of a bridge between the rainfall and infiltration processes. Interception is clearly associated with vegetation. The process is still the subject of experiments in nontropical areas and very limited information exist from the tropics, where the storm pattern is very special. Jackson [13] studied interception in an intermediate forest community, typical for tropical mountain slopes exposed to humid climates. The amount of interception as observed there during several storms is given in Tab. 3.2.

Tab. 3.2. Amount of gross rainfall and intercepted water as measured by Jackson in Tanzania

Storm size (mm)	0.7	1.5	1.6	2.9	3.5	3.6	4.7
Interception (mm)	0.6	1.5	0.9	1.5	2.7	0.4	2.6

Storm size (mm)	6.4	12.5	17.6	23.1	26.9	30.9	61.3
Interception (mm)	2.1	3.2	3.9	1.4	0.9	1.9	0.2

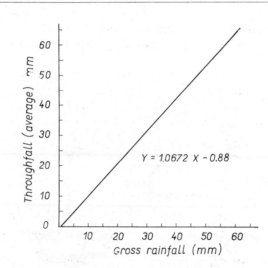

Fig. 3.10. Relationship between throughfall and gross rainfall as developed by Jackson in West Usambara Mountains, Tanzania.

The table indicates that the variability of the interception depends on the storm pattern. An additional relationship on throughfall and gross rainfall has been developed by Jackson (Fig. 3.10). Remarkably small values of stemflow were measured in tropical forest, less than 1 mm for instance, from storms of 40 — 65 mm rainflow. Tab. 3.3 contains smoothed values of the amount of interception from a gross rainfall.

Tab. 3.3. Smoothed values of intercepted water as depending on the gross rainfall

Gross rainfall (mm)	1	2.5	5	5.7	10	15	20	30	40
Intercepted (%)	80	44	28	21	17	12	10	7	5.5

Approximate estimates indicate that about 70—80% of the precipitation above the jungle reaches the soil surface, about 80% reaches the soil below the savanna grass and 80—85% below dense tropical crops.

According to observations from the Sudan [15] it has been concluded that precipitation, with a duration of 15 minutes, produces the same intensities in the tropics as in moderate regions, while a further increase in rainfall duration produces higher rainfall intensities in the tropics. This fact must be taken into account when future interception studies are made in the tropics. As described by Balek, Perry [8] the amount and intensity of rainfall, together with the maximum and instant storage capacity of the leaves, are decisive factors when the amount of water intercepted is to be calculated.

3.7 ROLE OF THE SOIL IN THE WATER BALANCE

Since tropical soils are discussed as a part of Chapter 4, only the role of the soil as related to the hydrological cycle is briefly discussed here. The function of the soil in a certain catchment is given by the prevailing soil type and thus in the tropics, where soil types vary greatly even in a small area, a detailed measurement of soil characteristics should be provided as a part of any water-balance study. The physical properties of the soils measured in an area less than 20 km² in size, situated 12° 34′ S, 28° 01′ E may serve as an example. Here six different units were localised (Tab. 3.5).

Similar variability was described by Pereira et al. [21] who studied two Kenyan experimental catchments (Tab. 3.4) on the flanks of volcanic mountain ranges.

Tab. 3.4. Soil moisture characteristics for the catchments Kericho and Kimakia, Kenya

	mm of water contained in 3 meters of depth at	
	Kericho	Kimakia
Field capacity (1/3 atm)	1450	1530
Wilting point (15 atm)	880	760
Storage capacity for water available to plants	558	770
Max. annual fluctuation in water storage	254	355
Range of annual streamflow	254—690	735—1470

Both results indicate that it is impossible to adopt any general conclusions on the role of the soil before an actual measurement has been provided.

Tab. 3.5. An example of the variability of soil types within a small area, 12°34′ S, 28°01′ E

Soil type:	63	63 S	83
Vegetation	Tall forest	Tall to open forest	Tall forest
Physiography	Catchment crest upper slopes, ridge tops	Middle of slope to intermittent swamp, sometimes upper slopes	Slopes to intermittent swamp of northern aspect
Relief	Nearly level	Nearly level to gently undulating	Sloping to gently undulating
Slope	1%	1—6%	3—4%
Erosion	None	None	None
Origin	Residual	Residual, upper layer may contain colluvial material	—
Plinthite and stone line	Frequent	Frequent	Characteristic
Depth (m)	1.5—3	0.5—1	0—1.3
Sand content 0.02—2.0 mm	85% sand in top 0.3 m, 70% at 1—3 m, increasing to 80% below	—	—
Dry season watertable depth (m)	6.5—13.0	6.5—8.0	3—6
Soil drainage	good	good	excessive
Permeability	moderate	moderate	rapid
Water content (mm/m)			
at saturation	130—140	—	—
at field cap.	75—100	—	—
at wilting p.	50— 65	—	—

3.8 EVAPOTRANSPIRATION IN THE HYDROLOGICAL CYCLE

This phenomenon is the most significant and most difficult factor to calculate in the water balance. The surface of the vegetation provides a temporary storage for rainfall water, as an interception. It is then lifted back to the air in a form which can be better described as evaporation than transpiration. The plants loose the water through transpiration through the leaf surface. Water is evaporated from the soil surface not covered by vegetation, and intermittent and perennial swamps, lakes and reservoirs also contribute to the evaporation. The main difficulty is that the total evapotranspiration cannot be measured with high accuracy and water

61 S	60	600
Open to close stunded, Footslopes bordering intermittent swamp	Open grass Footslopes bordering intermittent swamp	Open grass footslope at higher elevation
Nearly level to gently undulating	Nearly level to sloping	Gentle slope
2—6%	1—3%	1—5%
None	None	None
Residual, sometimes as 63 S	Residual. Possibly some celuvial/alluvial matter. in upper horizon	As for 60
Very frequent	Not common	Not common
0.5—1.7	1—2	1—2
85% sand top 0.3 m, 70% at 1—2 m, increasing to 80% below	95% in top 1 m, decreasing to 80% below	95% in top 1 m decreasing to 80% below
3—8	2.3—3	1—2.3
poor moderate to slow	poor slow	poor slow
210—340 210—340 85—210	120—465 210—465 40—175	— — —

balance calculations are still among the most efficient methods of evapotranspiration estimation.

Evapotranspiration calculation/estimates in the tropics have significantly contributed to the development of the whole field under study. For instance, estimates by Hurst [12], indicating that the Sudanese papyrus may evapotranspirate twice as much water as a free water surface, have formed a new view on the ratio of actual evapotranspiration to potential evaporation. Originally it was thought that the ratio cannot be higher than one. Similar results were gained by specialized measurements of evapotranspiration conducted in Uganda [8]. Measurements of the evaporation of the water hyacinth by Van den Weert and Kammerling [25] resulted in evapotranspiration values 3−4 times higher than the potential evaporation from a free water surface.

More species have been studied from the point of view of the daily process of transpiration. From measurements taken by Lawson and Jenik [16] on *Securinega virosa,* which is a typical shrub of the Accra Plain, significant values of evapotranspiration can be seen even during the night hours (Fig. 3.11). From repeated

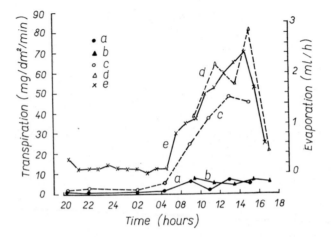

Fig. 3.11. Process of transpiration in Securinega virosa. a. leaves from windward side exposed on windward side, b. leaves from windward side exposed in standard position, c. leaves from leeward side exposed on the leeward side, d. leaves from leeward side exposed in standard position, measured 150 cm above ground in grassland, Accra Plains, Ghana.

experiments it followed that evaporation from the shrubs was lowest from 18.00 to 06.00 hours. It increased steadily and achieved a maximum between 12.00−14.00 hours. The authors concluded that the transpiration rate on the wind-exposed sides of the slump is much lower because of the higher desiccation of the soils. This is very likely valid only for certain periods of the year because water from the soil is mostly evaporated through the shrub. Temporary shading may also influence the microclimate of savanna species. A profile diagram of a strip 50 × 5 m plotted by Lawson indicates (Fig. 3.12) how favourable are the conditions for transpiration from various levels of shrubbed savanna. Several layers of leaves are exposed to the wind and radiation. Obviously, the higher the leaf area index, the higher the evapotranspiration, provided that the leaves are exposed to the radiation and wind for at least some part of the day.

Interesting results on evapotranspiration have been obtained by a detailed balance analysis of woodland, grassland and transitive areas. Tab. 3.6 give the values of monthly potential evaporation based either on sunshine or radiation records and

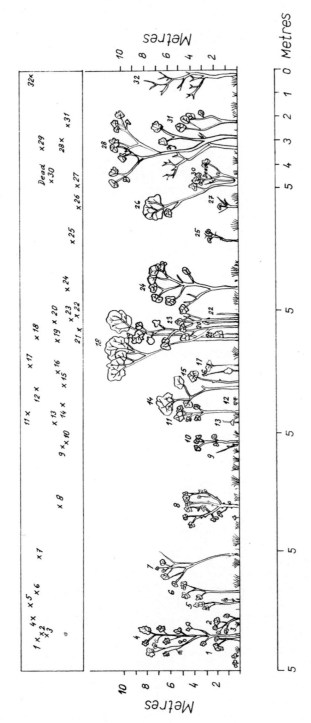

Fig. 3.12. Profile diagram in the middle-slope savanna, Ghana.

Tab. 3.6a. Monthly evapotranspiration from grassland, woodland and transitive zone, Central African Plateau, 1967/68

Unit	Precipitation	Potential evaporation		Evapotranspiration						
		Sunshine	Radiation	Grassland		Woodland		Transitive zone		
		E_0	E_0	E_t	E_t/E_0	E_t	E_t/E_0	E_t	E_t/E_0	
	mm	mm	mm	mm		mm		mm		
October	16.26	190.97		29.13	0.15	66.98	0.35	66.98	0.35	
November	152.91	171.11		124.00	0.72	152.73	0.89	152.73	0.89	
December	194.82	145.98		65.86	0.45	145.98	1.00	145.98	1.00	
January	344.68	146.42		99.64	0.68	146.42	1.00	146.42	1.00	
February	173.48	152.87		71.25	0.47	162.95	1.07	152.87	1.00	
March	258.32	161.63		55.30	0.34	197.69	1.22	158.98	0.98	
April	11.94	150.93		26.59	0.18	127.89	0.85	86.28	0.57	
May	0.00	130.22		10.34	0.08	87.45	0.67	51.00	0.39	
June	0.00	101.38		9.02	0.09	63.36	0.63	40.21	0.40	
July	0.00	114.75	113.66	6.65	0.06	61.38	0.53	42.11	0.37	
August	0.00	136.40	132.04	5.56	0.04	55.04	0.40	44.39	0.33	
September	0.76	159.87	151.63	5.13	0.03	46.71	0.29	45.26	0.28	
Year	1153.17	1762.53		508.47	0.29	1314.58	0.74	1133.21	0.64	

Tab. 3.6b. Monthly evapotranspiration from grassland, woodland and transitive zone. Central African Plateau, 1968/69

Unit	Precipitation	Potential evaporation		Evapotranspiration					
		Sunshine	Radiation	Grassland		Woodland		Transitive zone	
		E_0	E_0	E_t	E_t/E_0	E_t	E_t/E_0	E_t	E_t/E_0
	mm	mm	mm	mm		mm		mm	
October	2.03	207.73	194.01	6.58	0.03	49.05	0.24	49.05	0.24
November	199.39	156.16	148.17	80.72	0.53	85.39	0.56	85.39	0.56
December	484.63	129.04	131.48	86.36	0.66	129.04	0.99	129.04	0.99
January	399.54	153.16	145.56	87.50	0.59	153.16	1.03	153.16	1.03
February	201.17	124.01	123.44	67.54	0.55	163.75	1.32	115.32	0.93
March	247.65	140.45	132.29	64.14	0.47	193.66	1.42	131.80	0.96
April	80.77	143.92	133.66	29.26	0.21	192.72	1.39	128.01	0.92
May	5.08	121.77	119.68	20.40	0.17	119.82	0.99	54.10	0.45
June	0.00	101.69	97.97	9.17	0.09	82.47	0.83	36.40	0.36
July	0.00	109.58	106.59	9.63	0.09	72.49	0.67	37.49	0.35
August	0.00	134.36	126.97	6.65	0.05	68.09	0.52	39.95	0.31
September	0.00	159.84	156.30	5.84	0.04	56.23	0.36	40.44	0.26
Year	1620.26	1681.71	1616.12	473.79	0.29	1356.87	0.83	1000.15	0.61

Tab. 3.6c. Monthly evapotranspiration from grassland, woodland and transitive zone, Central African Plateau, 1969/70

Unit	Precipitation	Potential evaporation		Evapotranspiration						
		Sunshine	Radiation	Grassland		Woodland		Transitive zone		
		E_0	E_0	E_t	E_t/E_0	E_t	E_t/E_0	E_t	E_t/E_0	
	mm	mm	mm	mm		mm		mm		
October	16.26	190.97		29.13	0.15	66.98	0.35	66.98	0.35	
November	152.91	171.11		124.00	0.72	152.73	0.89	152.73	0.89	
December	194.82	145.98		65.86	0.45	145.98	1.00	145.98	1.00	
January	344.68	146.42		99.64	0.68	146.42	1.00	146.42	1.00	
February	173.48	152.87		71.25	0.47	162.95	1.07	152.87	1.00	
March	258.32	161.63		55.30	0.34	197.69	1.22	158.98	0.98	
April	11.94	150.93		26.59	0.18	127.89	0.85	86.28	0.57	
May	0.00	130.22		10.34	0.08	87.45	0.67	51.00	0.39	
June	0.00	101.38		9.02	0.09	63.36	0.63	40.21	0.40	
July	0.00	114.75	113.66	6.65	0.06	61.38	0.53	42.11	0.37	
August	0.00	136.40	132.04	5.56	0.04	55.04	0.40	44.39	0.33	
September	0.76	159.87	151.63	5.13	0.03	46.71	0.29	45.26	0.28	
Year	1153.17	1762.53		508.47	0.29	1314.58	0.74	1133.21	0.64	

the evapotranspiration values as obtained by water balancing. The results indicate that trees transpire three time more water than grass in the same climate. Evapotranspiration from the transitive zone varies according to the percentage of grass and trees [8]. The *Brachystegia* species, typical for great parts of the African plateau, can easily evaporate up to 90% of the total precipitation and the ratio can easily exceed 1.0 during the rainy season, in the woodland area.

Some studies of the water balance in the tropics presume a constant ratio E_t/E_0 in monthly and annual values. Considering the frequent fluctuation and availability of the groundwater level, soil moisture and precipitation pattern, it can be pressumed that the ratio also varies considerably.

When calculating the evapotranspiration for the whole catchment, it must be borne in mind that the ratio is variable even within the catchment limits. Infrared pictures of tropical catchments taken in different seasons of the year indicate this type of variability.

A fluctuation of evapotranspiration, depending on the vegetational cover and variability of annual rainfall, has also been calculated on a large scale. For instance, Balek [5] estimated the evapotranspiration for the river headwaters of the Upper Zambezi, in an area of 1500 km^2, at 1100 mm, for the middle Zambezi below the Barotse Flood Plains (drainage area 320,000 km^2) at 1000 mm and for the Zambezi basin to the Victoria Falls, including part of Kalahari (drainage area 1250,000 km^2), at only 600 mm. The decreasing evapotranspiration rate indicates a decreasing rate of rainfall which can meet the evapotranspiration requirements. The vegetational cover combined with rainfall pattern causes changes in the hydrological cycle.

In arid regions, it may seem that certain factors have a prevalent influence on evaporation. According to the experiments by Hellwig [11] near Windhoek with evaporation from moist sand, evaporation was highest just before sunrise. This has been laid to the account of the difference between the temperature at the water table and that of the air, in the absence of the influence of radiation.

3.9 A CONCEPTUAL MODEL OF THE WATER BALANCE IN A TROPICAL BASIN

In an attempt to study the water balance of tropical catchments and its fluctuation within short time intervals, a general model of the tropical water balance has been set up [7]. The main purpose was to describe the water storage and its fluctuation in the main parts of the hydrological cycle. A flow chart of the model is found in Appendix 1. The name dambo indicates that the model involves the simulation of the hydrological regime in an intermittent swamp, called in some parts of Africa dambo. The list of the symbols used is found in Appendix 2.

A general tropical catchment has been described as consisting of two regions, deep and shallow. The shallow region corresponds approximately to the alluvial part of the basin or the part of rapid groundwater level response the rainfall regime. Intermittent or perennial swamps may also form part of it. More regions, called transitive regions, can be found in some catchments and the model makes it possible to extend the number of regions if necessary.

There are three main zones in each region. First, the groundwater zone is defined as a noncapillary pore space between the groundwater level, corresponding to zero outflow, and the soil surface. The zone can be overdepleted and then the instant groundwater storage appears as negative. This is typical for a situation where the groundwater outflow has diminished and the aquifer is still being emptied by evapotranspiration. Thus, the possibility of intermittency, an important phenomenon in many tropical basins, is involved in the model.

The zone of capillary rise is associated with the groundwater zone and it is recharged either from the groundwater zone or from precipitation. Actually, the replenishment from the groundwater zone is much more intensive in the tropics, because the zone of capillary rise is a permanent source of water for many plants during the prolonged dry season. Precipitation can recharge this zone up to a certain maximum, sometimes still known as the field capacity, otherwise the excess contributes toward a rise in the groundwater level. The zone can be recharged up to a certain stage from the groundwater sources in accordance with the pF curve, which is characteristic for a given soil profile. The zone is depleted by evapotranspiration only up to a certain limit, more or less corresponding with the so-called wilting point. Both terms, field capacity and wilting point, have been widely criticised, but they are still used in the model as characterizing certain critical stages of soil moisture, determined by trial and error and not necessarily corresponding with the standard definition.

The upper moisture zone is located between the soil surface and the zone of capillary rise. It can be recharged only by precipitation or, indirectly, as a space of capillary pores filled with water after decrease in the groundwater table and followed by decrease in the zone of capillary rise. Obviously, the zone of capillary rise can be recharged in such an indirect way as well. When the groundwater zone decreases, both soil zones increase and vice versa. The upper moisture zone can be depleted by evapotranspiration. Both soil moisture zones can diminish partially or completely when the groundwater level rises. However, the water storage of both zones is then hidden in the capillary pores of the aquifer.

The vegetation and soil surface form a so called surface-biological zone where the water is intercepted during rainfall and released either for infiltration into the soil or for evapotranspiration. The precipitation rate which exceeds the interception rate, forms the surface runoff which can be further increased by the excess of the groundwater storage and this may result in the formation of intermittent swamps.

Elsewhere in the tropics, a significant seasonal variability of leaf area and active root depth can be expected. While the first factor influences the size of the interception storage capacity and the instant magnitude of the infiltration rate into the surface-biological zone, the second must be estimated as a basis for deciding whether or not the roots are in contact with the zone of capillary rise. The formula

$$C_b = \frac{C_{by} - C_{bx}}{2} \left\{ \sin 0.986 \left[x - (B - 91) \right] \right\} + \frac{C_{by} + C_{bx}}{2}$$

determines the seasonal fluctuation. Here C_b represents the current capacity of the surface-biological zone, root depth or interception rate, C_{by}/C_{bx} is the maximum/minimum storage capacity, root depth or interception rate and represents the corresponding minimum values. B is the calendar number of the day on which one may expect the highest seasonal development of each particular factor.

The interception rate is determined by using this formula in the event that no water has been stored in the biological zone of the catchment, otherwise, a reduced interception rate must be determined

$$f_i = \left[1 - \left(\frac{\log \frac{W_b}{f_m}}{\log \frac{C_b}{f_m}} \right)^2 \right] f_m$$

where W_b is the current storage of intercepted water, C_b is the current capacity of the surface biological zone and f_m is the maximum interception rate which may exist within a year. Obviously the condition

$$C_b \geqq W_b > f_m > 0$$

must be satisfied, otherwise the rate needs to be defined by special statements.

Evapotranspiration is probably the most decisive factor in water balance calculations. Penman's formula has been widely used in the tropics for the calculation of potential evaporation. In attempting to find a bridge between the potential evaporation and actual evapotranspiration, all possibilities for the water to transpire/evaporate have to be considered. The first opportunity for water to be evaporated is from the surface of the vegetation or, in other words, from the intercepted water. An amount of water equal to potential evaporation and related to the unit area, should be used in the first trials of the model. The more significant part is taken from the subsurface horizons. The upper moisture zone satisfies transpiration needs at first. It must then be decided whether the roots are in contact with the zone of capillary rise, and how much of the transpiration is taken from there. Referring to the previous discussion on root systems, water is taken from both zones simultaneously, provided that the roots are developed sufficiently deeply. The rate of transpiration depends upon the instant water storage and the evapotranspiration decreases with decreasing soil moisture:

$$E_{tu} = E_0 \frac{S_u - S_{uw}}{S_{uf} - S_{uw}}$$

Here E_{tu} is evapotranspiration from the upper moisture zone, E_0 is potential evaporation, or that part of it which has not been previously saturated, S_u is the current soil moisture in the upper zone, S_{uw} is the soil moisture corresponding to the wilting point and S_{uf} is the soil moisture content corresponding to the field capacity. Obviously, when the soil moisture is below S_{uw}, there is no evapotranspiration from the upper moisture zone.

In tropical catchments, the roots are usually within reach of the zone of capillary rise. If the roots are not able to cover E_{tu}, an additional supply can be provided from the zone of capillary rise. The following formula after Rijtema [23] has been adapted for the model

$$E_{tl} = \left(\frac{A}{H_d}\right)^{\frac{1}{B}}$$

where E_{tl} is evapotranspiration from the zone of capillary rise, H_d is the groundwater depth and A and B are coefficients depending upon the soil type. Supposing the zone of capillary rise is not fully saturated, the evapotranspiration rate from there is reduced so that

$$E_{tl} = \frac{S_1 - S_{lw}}{S_{lf} - S_{lw}} \left(\frac{A}{H_d}\right)^{\frac{1}{B}}$$

where S_1 is the instant moisture in the zone of capillary rise, S_{lw} is the minimum moisture storage to which the capillary rise zone can be depleted and S_{lf} is the soil moisture corresponding with the fully saturated stage.

When the groundwater level reaches the surface, the evaporation from the soil reaches its maximum.

The groundwater zone is saturated through the upper and lower moisture zone after both have become fully saturated. It is emptied by a direct uptake of water into the zone of capillary rise, recharging the transpired water, with and intensity estimated by the trial and error method. Cooperation between a hydrologist and a soil physicist is highly recommended at this point. Although the groundwater zone is depleted by the groundwater outflow, this seems to be less significant in the tropics than depletion by evapotranspiration. A simple formula has been used to characterize the groundwater outflow

$$q_i = q_0 K^t$$

where q_i is the final and q_0 the initial discharge value, t is the time interval between them and K is recession coefficient characteristic for the given aquifer and time unit used. Then the groundwater storage can be determined as

$$W_g = \frac{Mq_i}{A \ln K}$$

where W_g is the groundwater storage, M is the constant depending on the time interval, q_i is the instant discharge, A is the drainage area of the balanced region.

As can be seen from App. 1, the relation between effective rainfall calculated by water balancing and the surface runoff hydrograph is calculated by using the linear differential equation of the second order

$$C \frac{d^2 y}{dt^2} + B \frac{dy}{dt} + Ay = x(t)$$

where $x(t)$ is the function of effective rainfall, y the response curve and A, B, C are the coefficients of the differential equation estimated by using special graphs (Ayers, Balek 1967). Obviously, any other equation may replace the previous one, should it prove to be more efficient. The same can also be said of any other formula previously mentioned in this Chapter, because the concept of the model allows the relatively simple replacement of any part.

3.10 WATER BALANCE CALCULATED FROM BASIC DATA

The previously described model may be found to be a useful tool in catchments which are fully controlled and with a dense observational network. Unfortunately this type of catchment is rather exceptional in the tropics and so in many cases the water balance should be based on monthly or annual values of rainfall and runoff. In this case, an intensive field survey of the whole basin should be provided, together with the rainfall-runoff relationships based upon long-term obervation over at least a decade and, if possible verified by at least a temporary dense network. A comparison should be made with long-term records, which may exist in the region, to discover whether the analysed data repesent a wet, dry or normal period. Data used for balancing should be chosen from the same period of observation, even if it means eliminating some longer records. Supplementary statistical and analytical methods should be used where a single, normally successful method has failed. For instance, when an analysis of monthly rqinfall − runoff data is made, the correlation coefficient is often found to be very low. This can be partly improved by using a multiple correlation analysis which involves the influence of previous months. The relationship between monthly rainfall and runoff calculated for the basin of the river Pra in Ghana can serve as an example, with the aim of extending the monthly discharge sequence by using long-term records of monthly precipitation. While a simple

78

regression relationship failed to provide reasonable data for eight months of the year, a better result was achieved by using the precipitation records from previous months (Tab. 3.7).

Tab. 3.7. Influence of precipitation on the monthly discharge of river Pra in Ghana

Monthly discharge in	Influenced by precipitation from
January	November, December
February	December, January, February
March	No significant relationship
April	No significant relationship
June	No significant relationship
July	April, May, June, July
August	July, August
September	August, September
October	July, August, September, October
November	August, September, October, November
December	September, October, November

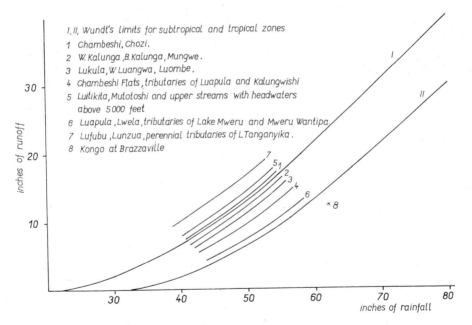

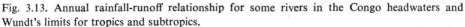

Fig. 3.13. Annual rainfall-runoff relationship for some rivers in the Congo headwaters and Wundt's limits for tropics and subtropics.

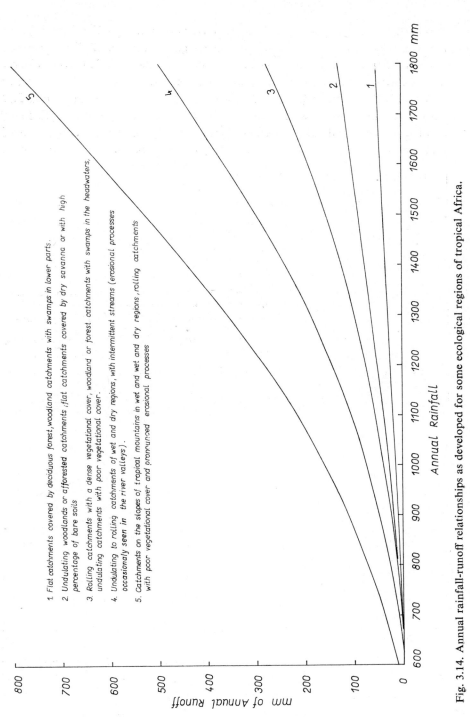

1. Flat catchments covered by deciduous forest, woodland catchments with swamps in lower parts.

2. Undulating woodlands or afforested catchments; flat catchments covered by dry savanna or with high percentage of bare soils

3. Rolling catchments with a dense vegetational cover, woodland or forest catchments with swamps in the headwaters, undulating catchments with poor vegetational cover.

4. Undulating to rolling catchments of wet and dry regions, with intermittent streams (erosional processes occasionaly seen in the river valleys).

5. Catchments on the slopes of tropical mountains in wet and wet and dry regions, rolling catchments with poor vegetational cover and pronounced erosional processes

Fig. 3.14. Annual rainfall-runoff relationships as developed for some ecological regions of tropical Africa.

The annual rainfall runoff relationship is usually plotted in the form a graph representing a typical river or region. By using a graph, the runoff of an ungauged basin can be determined provided the annual rainfall of the basin is known. If there are no data available at all, then the guiding curves as developed by Wundt can be used. Wundt's limits for tropical and subtropical regions are plotted in Fig. 3.13 together with the graphs drawn for the rivers of the Upper Congo sysetm [3]. Many rivers of the tropics and subtropics may behave outside the limits, as was proved experimentally. Estimates based on these curves should be taken as approximate.

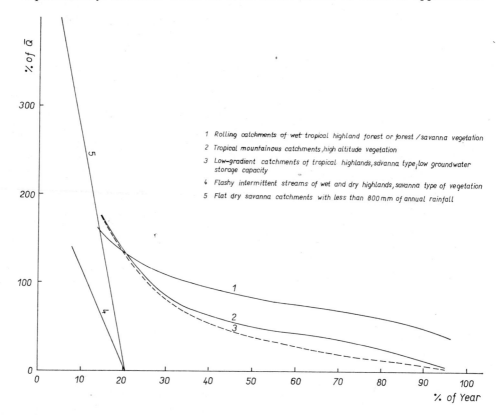

Fig. 3.15. Characteristic duration curves as developed for some ecological regions of tropical Africa.

In an attempt to provide more general graphs, a diagram has been sketched showing the set of curves for basins which can be characterized by certain parameters [8]. The relationships have been developed by analysing the data from six experimental catchments and numerous streams in the basins of lakes Victoria, Albert and Kyoga (Fig. 3.14). There are five different categories among the catchments according to the slope of the basin, vegetational cover and hydroecological classification. When using such a graph it should be borne in mind that very few streams behave along their whole course in accordance with a single curve. Radical changes of the basin slope, vegetational changes, or the occurrence of swampy areas, may form a special graph fluctuating either smoothly or in jumps from one guiding curve to another.

Other significant characteristics in water balancing are the duration curves. A family of curves has been developed as a tool when no information exists on the distribution of runoff in a year (Fig. 3.15). The curves at the lowest and highest values should be used very carefully. These values may vary greatly from stream to stream and from year to year, even within a small area. As an example of a water-balance calculation made by using basic data, the values for the Congo headwaters are given in Tab. 3.8 [4].

Sometimes designers of water-management projects require not only the percentual distribution of the runoff in a year, but also distribution on a monthly basis. Elsewhere in Africa, the monthly distribution of the rainfall is usually known. By using the runoff distribution of any intensively observed catchment at which the precipitation distribution is similar to the studied area, the monthly runoff a as percentage of the mean annual runoff can be estimated. A graph is plotted in Fig. 3.16, as an

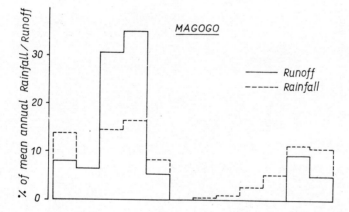

Fig. 3.16. Monthly distribution of rainfall and runoff at Magogo, Tanzania.

example of the monthly distribution of mean annual runoff. Both precipitation and runoff values are plotted as percentages of the relevant mean annual values. According to the graph, the intermittency of the streams within the area is estimated for five months, although precipitation occurs during eleven months of the year.

Separation of various components of river flow also belongs to the problems which are studied as a part of water balance, because such an analysis provides useful information on the hydrological cycle in a given area.

Tab. 3.8. Water balance of some parts of the Upper Chambeshi basin, Congo headwaters

Upper Chambeshi Water Balance	Drainage area	Mean annual rainfall	Mean annual runoff	Mean annual loss	Runoff coeff.	Water yield	Mean annual disch.	Discharge likely to be excessed				
								20	40	60	80	95
								% of a year				
Unit	m²	in	in	in	%	cfs/m²	cfs	cfs				
Chambeshi CG8	1580	46.0	10.70	36.30	23.3	0.789	1245	2855	1370	560	162	31
Interbasin Chambeshi GG8 – Wiwa Kalungu	673	42.0	8.33	33.67	19.8	0.613	413	965	455	190	53	10
Chambeshi above Wiwa Kalungu	2253	44.8	10.00	37.80	22.3	0.756	1658	3820	1825	740	215	41
Chozi CG7	849	43.1	8.70	34.40	20.2	0.640	544	1250	600	245	70	14
Interbasin CG7 – Chozi mouth –	1215	39.9	7.00	32.90	17.5	0.515	626	1440	690	255	80	15
Wiwa Kalungu	2064	41.2	7.70	33.50	18.7	0.566	1170	2690	1290	500	150	29
Wiwa Kalungu CG12	1121	40.0	5.40	34.60	13.5	0.397	446	865	245	105	30	5

Rem.: The observed and gauged cross sections are underlined.

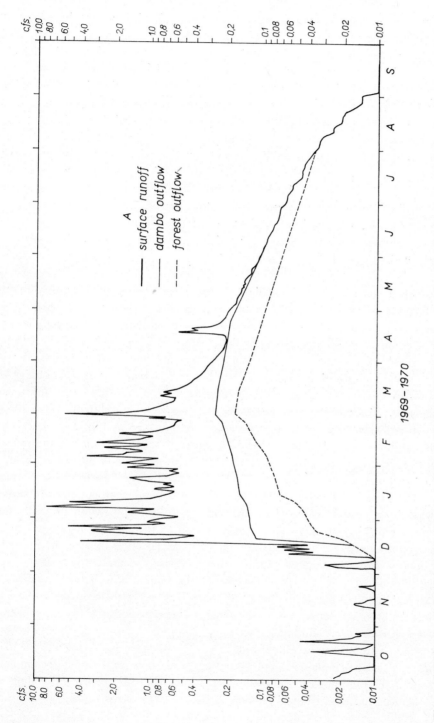

Fig. 3.17. Three-components separation of annual hydrograph. Catchment A at the Luano catchments, Zambia, 1969—1970.

Separation of various components of river flow also belongs to the problems which are studied as a part of water balance, because such an analysis provides useful information on the hydrological cycle in a given area. Fig. 3.17 depicts a separation based on the results of the model described earlier in this Chapter. The balance has been calculated for a small intermittent stream with a drainage area of 1.43 km². We can see from the separation that the contribution of surface runoff continues for approximately fifty days after the last flood peak has occurred. This means that only when the interval between two storms is longer then fifty days, does the hydrograph follow a pure groundwater outflow. As such a situation occurs rarely during the rainy season in a given area, the hydrograph is permanently formed by the surface runoff component. A period of fifty days may appear to be too long for such a small catchment when compared with other catchments of much greater size. The explanation is found in the high resistance of the grass to the surface flow.

It can be traced from the same picture that when the discharge falls below say 0.02 cfs, the hydrograph shape becomes influenced by evaporation from the catchment.

The hydrograph has been separated into three components, surface runoff, outflow from the woodland and outflow from the grassland. While the surface runoff diminishes every year under wet and dry climate, the groundwater outflow may or may not diminish. The diagram shows that there is still an outflow at the end of September and it is formed by the groundwater component. Surface runoff contributes at the beginning of the rainy season, but is still very low. The discharge is influenced by evaporation until the rains saturate the soil moisture zones and form another active groundwater storage. When the stream is intermittent, the shallow region groundwater zone is recharged earlier, as indicated in Fig. 3.18; later on, there is a groundwater outflow from the deep region. When the stream is not intermittent, the groundwater outflow is saturated from the deep region at the end of the dry and the beginning of the wet season, as in Fig. 3.17, and an outflow from the shallow region starts later. The peak of the deep region groundwater component varies somewhat from year to year, while the peak of the shallow region groundwater component remains constant year by year, owing to the limited storage capacity of the groundwater zone in the shallow region. The depletion time of the groundwater outflow from the shallow region is therefore also constant from year to year and the intermittency of the stream depends on the amount of water accumulated in the deep region during the wet period of the same year and of the previous year.

When the groundwater storage in the shallow region exceeds the storage capacity of the groundwater zone, another surface runoff component is formed. This, however, is not common to all tropical catchments.

Thus, by using simple evaluation routines and advanced model techniques, a picture of the hydrological behaviour of the catchment is obtained. Hydrological water balance studies in tropical regions, in fact, in many cases supply more infor-

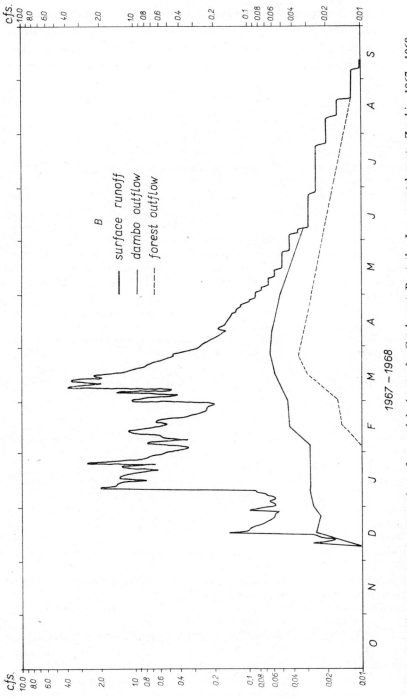

Fig. 3.18. Three-components separation of annual hydrograph. Catchment B at the Luano catchments, Zambia, 1967—1968.

mation, even with limited data, then studies in temperate regions because of the more uniform fluctuation of wet and dry seasons, absence of snowfall, soil freezing and thawing and, last but not least, because the tropical catchments are still far less influenced by man's activities.

3.11 LIST OF LITERATURE

[1] Aubréville, A. et al., 1959. Vegetation map of Africa south of the Tropic of Cancer. Clavendon Press, Oxford.

[2] Ayers, H., Balek, J., 1967. Derivation of a general flood wave hydrograph from a continuous water balance. UNESCO Symp. on Floods, Leningrad, 14 p.

[3] Balek, J., 1970. Water balance of Luapula and Lake Tanganyika basins, WR 6 Rep. N. C. S. R. Lusaka, 15 p.

[4] Balek, J., 1971. Hydrological data for Upper Zambezi and Upper Congo Headwaters. Symp. on the role of hydrol. in Africa, Addis Ababa, 1971, 10 p.

[5] Balek, J., 1971. Water balance of the Zambezi basin. WR 8 N. C. S. R. Lusaka, 20 p.

[6] Balek, J., 1973. Use of experimental and representative catchments for the assessment of hydrological data of African tropical basins. UNESCO Symp. on water res. projects with inadequate data, Madrid, 11 p.

[7] Balek, J., Perry, J., 1972. Luano catchments, first phase final report. WR 20 N.C.S.R. Lusaka, 69 p.

[8] Balek, J., 1972. An evaluation of the index catchments and extension of the data to the ungauged areas in the hydrol. survey of lakes Victora, Kyoga and Albert. Interim report for W. M. O., Entebbe, 29 p.

[9] Beard, J. S., 1964. Savanna. In the "Ecology of man in a tropical environment", Mogens, Switzerland, 10 p.

[10] Dagg, M., Pratt, M. A. C., 1961. Stormflow from a forested catchment. Hydrol. symp. Nairobi, 8 p.

[11] Hellwig, J., 1973. Evaporation of water from sand, diurnal variation. *Journal of Hydrology* **18**, pp. 109—118.

[12] Hurst, H. E., 1954, Le Nil. Payot, Paris.

[13] Jackson, I. J., 1971. Problem of throughfall and interception assessment under tropical forest. *Journal of Hydrology,* **12**, pp. 235—253.

[14] Kimble, G. T., 1960. Tropical Africa. Twentieth Cent. Found., N. York.

[15] Kutilek, M., 1963. Tropical and subtropical soils. Prague Tech. Univ. Press, 109 p. (in Czech).

[16] Lawson, G. W., Jenik, J., 1967. Microclimate and vegetation in the Accra Plains. J. Ecol. 55, 777—785.

[17] Lawson, G. W., et al., 1968. A study of a vegetation catena in Guinean savanna and Mole game reserve (Ghana), 1968. J. Ecol. 56 pp. 505—522.

[18] Matějka, V., 1971. Forest management in tropics and subtropics. Prague Agr. Univ. Press, 307 p. (in Czech).

[19] Maxwell, D., 1972. Root range investigation. TR 24 Rep. N. C. S. R. Lusaka 15 p.

[20] Pereira, H. G., 1962. Hydrological effects of changes in land use in some East African catchment areas. *East Af. Agr. For. Journ.* **24**.

[21] Pereira, H. G., Dagg, M., Hosegood, P. H., 1961. Intensive method of catchment basin study. Hydrol. Symp. Nairobi, pp. 325—337.

[22] Pollock, N. C., 1968. Africa. University of London Press.

[23] Rijtema, P. E., 1966. Analysis of actual evapotranspiration. Pudoc, Wageningen.

[24] Walter, H., 1964. Productivity of vegetation in arid countries. Ecology of man in trop. environ. Mogens, Switzerland, 9 p.

[25] Weert Van der, R., Kamerling, G. E., 1974. Evapotranspiration of water hyacinth. *Journal of Hydrology* **22**, pp. 201—212.

[26] Whyte, R. O., 1966. The use of arid and semiarid lands. Unesco symp. Arid Lands.

[27] Wicht, C. C., 1949. Forestry and water supplies in South Africa. *Dep. Agr. S. Afr. Bull.* **58**, 33 p.

[28] Worthington, E. B., 1964. Inland waters. In Ecol. of man in trop. envir., 7 p.

[29] Wundt, W., 1953. Gewässerkunde. Heidelberg.

4. AFRICAN RIVERS

4.1 BASINS AND DIVIDES

A very typical pattern of African relief is a broad shallow basin, separated by the divides formed by fault blocks, mountain ranges and plateaux, where rock wastes eroded from the plateau surface have been deposited in the basin. The most significant basins of this type are the Congo, Chad, El Dhouf, Kalahari and Victoria-Kyoga.

Fig. 4.1. River network between Lake Victoria and George influenced by the backwash of the rivers.

Fig. 4.2. Catchment of the Upper Niger, according to Pritchard.

A striking feature of these basins is the direction of the drainage system. Instead of following the prevailing westward slope of the plateau, the networks follow different directions with the exception of the Orange river. Thus, the Niger flows north-east and south, the Congo north and southwestwards, the Nile almost north and the Zambezi to the east. Some geologists assume that the draingae network was formed during the period of existence of the so-called Gondwanaland, a supercontinent on which Africa was inland. After the disintegration of the supercontinent, the rivers, with few exceptions, found their way to the sea. In some areas, particularly in Uganda, the effect of the backtilting still can be traced. Here the backwash of the rivers along their courses has formed irregular marshy lakes and altered the flow patterns of many rivers. This effect is found particularly in the area between Lake Victoria and Lakes George and Edward (Fig. 4.1). As another example, Fig. 4.2 gives the capture of the Upper Niger drainage network.

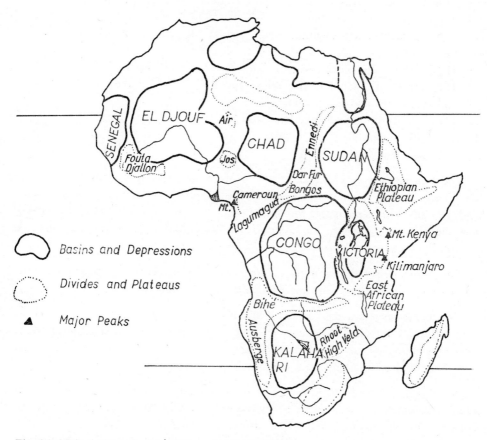

Fig. 4.3. Main drainage basins/systems of tropical Africa.

The classification of the main drainage basins of the African tropics is quite simple; the various attempts to differentiate between the basins are usually nonuniform, particularly where the short coastal basins are concerned. They are often thought to be joint zones between the mouths of the main tropical streams. Fifteen areas can be traced by using this classification (Fig. 4. 3):

1. The Senegal river basin.
2. The West African coast between Gambia and Comoe and Bandama basins.
3. The Volta basin.
4. The Niger/Benue basin.
5. The Lake Chad basin.
6. The Chari/Lagone basins.
7. The West African coast between the Sanaga and Ogoué basins.
8. The Congo drainage system, including the Chambeshi, Luapula, Uele, Ubangi, Sangha, Lukene, Kasai and Kwango rivers and Lake Tanganyika.
9. The Limpopo and Save basins.
10. The Zambezi system including the Cubango, Kafue and Luangwa rivers and Lake Nyasa.
11. The East African coastal basins south of Suez to the Lurio basin.
12. The Nile basin, including the White and Blue Nile and the Victoria, Kyoga, Albert, George and Edward Lakes.
13. The Lake Rudolph and Lake Eyasi basins.
14. The Etosha pan.
15. The Lake Rukwa basin.

The Senegal river is a confluence of the rivers Bafing and Bakoye flowing from the Fouta Djallon Highland in Guinea. The river is approximately 1100 km long and only part of it is navigable for small boats. The river forms the Mauritania-Mali and the Mauritania-Senegal boundaries. The lower part of the river is connected with Lakes Guiers and Rkiz.

The Gambia river drains parts of the Fouta Djallon Highland to the Atlantic. The river is over 1000 km long and navigable up to the Barrakuda falls.

The Volta river is a confluence of the Black, White and Red Volta and flows southward into the Gulf of Guinea. The river is navigable only for sixty miles. Many upstream rapids are now flooded by the reservoir Volta Lake, the dam of which has been built eighty kilometres from the mouth of the rivers.

The Niger river rises in the Fouta Djallon Highland and flows 4100 km into the Gulf of Benin. Above Timbuktu the river is called Joliba. At Segou, it flows into flat country, flooding almost 80,000 km² of the inland delta. At Timbuktu the river changes its the direction to the east. Important tributaries below Gao are the Sokoto, Kaduma and Benue. The delta begins 130 kilometres from the coast. The delta channels are known as Oil rivers. The Benue river, 1400 km long, is the largest tributary of the Niger. Its main tributary, the Manyo-Kebbi, is reported to be connected occasionally, during the flood period, through the Logone with Lake Chad. 800 kilometres of the Benue are navigable; important tributaries are the Faro, Katsion, Ala, Gongola and Donga.

The Lake Chad system is sometimes thought to be one system with two main tributaries, the Logone and Chari. There is no permanent outlet from the lake, but an overflow was reported into the Bahr el Ghazal basin in 1956 after several years of lake rise. The normal size of the lake is 32 000 km², the size of the surrounding swamps fluctuates between 10,000 and 25,000 km². The water is brackish.

The West African coast between Sanaga and Ogoué forms another hydrological unit. A few small rivers south of Ogoué also belong to the area. The direction of the streams is westward; they are rather short because the eastern slopes of the head-waters are drained into the Congo system.

The Congo basin is the largest in Africa, second only to the Amazon on a world scale. The length of the river is estimated at 5100 km. The source of the river is thought to be the Chambeshi river rising south of the Lake Tanganyika and transformed by the Bangweulu Swamps into the Luapula river. After flowing through Lake Mweru, it joins the Lualaba. Above the confluence, the Lualaba passes through Lake Kisale and is joined by the Lufira. Lake Tanganyika joins the Congo system through the Lukuga river. The most significant tributaries of the Congo are the Lomami, Aruwimi, Itimbiri, Mongala, Lulonga, Momboyo, Ubangi and the drainage of Lake Tumba. Another strong tributary, the Kasai, is formed by the rivers Sankuru, Lulua, Lushiko, Kwilu/Kwa. There are numerous rapids and cataracts in the system. Hell's Gate and Stanley Falls divide the upper and middle navigable parts. Below them, the river is navigable for 1600 kilometres. Near the ocean, the river flows through a narrow gorge and thirty-two cataracts, known as the Livingstone Falls, divide the navigable stream from the Atlantic.

Rather small rivers form the hydrographical system between the Cuanza and Cunene rivers. They rise on the Bié Plateau of Angola and flow to the Atlantic. Parts of both rivers are navigable.

South of the preceding system is the *Etosha Pan,* a closed system with no outlet. Several intermittent streams carry water into the pan.

The most southern basin of the African tropics draining into the Indian ocean is that of the *Limpopo or Crocodile river.* The river rises in the vicinity of the Indian ocean on the north side of the Witwatersrand and flows in a semicircular course, 1700 kilometres long, into the ocean. The many tributaries are only intermittent. The most significant tributaries are the Shashi and Olifants. Below the Olifants river the Limpopo is navigable.

The Zambezi is the largest African stream flowing into the Indian ocean. The 2600 km long river rises on the Central African Plateau and flows southward. It is joined by the Kabompo, Lungwebungu and Luanginga. Before joining the Chobe system from Northern Kalahari, it turns eastward. The system of the Chobe is formed by the rivers Kwando and Okawango and it rather complicates the hydrography of the Zambezi, since connection between the Chobe and Zambezi can occur in either direction and the Zambezi drainage area below Chobe can be thus taken to be somewhere between 500,000 and 1200,000 km^2.

The Victoria Falls below Chobe are one of the largest in the world. They are 1650 meters wide with a mean annual discharge of 1237 m^3/s falling to a depth of 98 meters. The lower part of the river regime is influenced by the Kariba and Caborra Bassa dams. Bellow Kariba, the rivers Kafue, Luangwa and Shire, draining the Nyasa Lake, join the Zambezi.

A strip of tropical Africa, south of Suez to the mouth of the Zambezi along the Red Sea and the Indian Ocean, is drained by short streams of various hydrologic regimes. The Awash, for instance, rises near Addis Ababa on the Ethiopian Plateau and, although rich in water upstream, eventually disappears in the middle of the Danakil desert. Generally, the streams can be characterized as short and frequently intermittent for a relatively large period of the year.

The Nile system is one of the most important in the tropics. The Nile itself is the longest rivers in the world, with a length of 6671 km. It flows from the equatorial plateau of East Africa, through the tropical plains of the Sudan, to the deserts of Sudan and Egypt. Victoria Lake used to be considered the main source, although the real spring of the river can be traced to the headwaters of the Kagera river, the largest tributary of the lake. The lake outlet is now controlled by a dam at a place once called the Rippon Falls. After passing Lake Kyoga, the Nile leaves the equatorial plateau at the Murchinson Falls, entering the Albert Lake. At its outlet, the river is known as the Albert Nile, downstream as the Bahr el Jebel. After being joined by the Bahr el Ghazal and Sobat, the river is called the White Nile, according to some sources because of the lack of river sediments. At Khartoum, the river is joined by the Blue Nile, rising in Lake Tana. The river Abbai, flowing into Lake Tana is, perhaps more correctly, regarded as the true source of the Blue Nile. From the confluence at Khartoum for the next 3100 kilometres, only the river Atbara, flowing from the Ethiopian Plateau, joins the Nile. Downstream, the river regime is under the influence of a system of reservoirs and irrigation channels. The Aswan Dam, with an overyear effect is most significant, but is already beyond tropical limits. A simplified scheme is given by Shahin in Fig. 4.4. The area of the inland lakes, having no connection with the ocean, forms a rather narrow strip between the Nile basin and the Eastern coastal basins. The lakes are of various size and have a typical high ratio of length to width. Three main pans are recognized between the Ethiopian Highland and Tanzania, namely the basins of Lake Rudolph and Eyasi and the Lake Rukwa basin. Their hydrological regimes depend mainly on their elevation above sea level and their latitude, as factors influencing the evaporation from the lakes. Lake Rudolph, fed by the river Omo flowing from the Ethiopian Highland, is the most important. It is 300 kilometres long and only 42 kilometres wide on the average. Its greatest depth is only 75 metres and its few minor tributaries are intermittent.

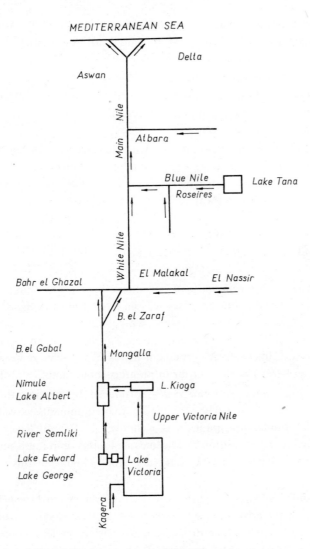

Fig. 4.4. Drainage scheme of the Nile system.

4.2 CLASSIFICATION OF TROPICAL RIVERS

In contrast to geographical classification, hydrological classification is completely different. First of all, it is closely related to the climate. In one of the very first attempts, Voeykov [30] characterized tropical rivers as being fed by rainwater and being at their highest during the summer season. Pardé [15] recognized:

a) tropical rainfall regimes characterized by a minimum discharge during the winter season and a maximum during the summer months (July, August, September, north of the equator and February, March and April, south of it).

b) equatorial types with two peaks.

Lvovich [19] differentiated in the tropics between

a) rivers fed only during autumn (such as the Congo and the White Nile),

b) rivers fed mostly during autumn (Niger, Nile),

c) rivers fed during summer.

Kimble [16], stated in 1960 that: "... with the exception of the Nile, the African rivers are still pretty much unknown". Since then, however, much knowledge has been gained. Nevertheless, it is true that the duration of observation of the majority of African rivers is still too short for more definite conclusions.

Contrary to the simplified classifications, a much greater variety of regimes is found between the two extremes represented, on the one side, by the tropical equatorial rivers and, on the other, by the ephemeral wadís. The following are the main river types found in tropical Africa:

1a. *Equatorial rivers with one peak,* produced by a heavy annual precipitation of 1750 – 2500 mm without a marked dry season. Even if most of the observation points in the basins record two periods of increased precipitation, one peak is much higher corresponding with the peak of precipitation. In Fig. 4.5 is an example of the river Sanaga at Edéa, Cameroon. A permanent increase in discharge between March and October is seen here. The ratio of mean monthly minimum/maximum is 0.164.

1b. *Equatorial rivers with two peaks,* produced by precipitation regimes with monthly totals over 200 mm. The catchments are predominantly covered by equatorial forest and the annual precipitation is well over 1750 mm. River Ogoué, Gabon (Fig. 4.5) is given as an example. The mean monthly minimum/maximum ratio is 0.250. In this particular case, the peaks are equally high, but this is not always the rule.

2a. Wet and dry types of climate in the lowland produce *tropical wet and dry lowland rivers,* in the areas covered by the moist type of savanna with a pronounced seasonal effect of rainfall. The dry season persists for at least three months. River Mangkoky (Fig. 4.5) in the west part of Malagasy serves as an example of such a river type. Although there are seven dry months, in the basin some precipitation always falls during this period. The decrease toward the end of the dry season is rather steep and thus the ratio of mean monthly minimum/maximum is only 0.053.

2b. *Rivers of the tropical wet and dry highlands of a moist type,* have their basins covered by woodland and savanna of a relatively moist type. The length of the dry season varies from place to place. The Zambezi headwaters, for example have more than 1400 mm of annual rainfall, while the tributaries of the middle part receive only 700 mm. The Upper Zambezi at Chavuma Falls (Fig. 4.5) is given as an example. The dry season is shorter than in the previous example, though there are at least

Fig. 4.5. Main river types: Sanaga river, Cameroon. Ogoué river, Gabon. Mangkoky river, Malagasy. Upper Zambezi, Zambia. Gwaai river, Southern Rhodesia. Pangani river, Tanzania.

96

three months without any rainfall. The mean monthly minimum/ maximum is 0.033, and clearly reflects the annual rainfall pattern.

2c. *Rivers of the tropical wet and dry highlands of a relatively dry type.* The basins are found in the marginal parts of the dry climate zones. In general, the basins receive little more than 500 – 700 mm, which is typical for semideserts. The basins are predominantly covered by dry savanna. River Gwaai, Southern Rhodesia (Fig. 4.5) is given as an example. Although the drainage area is almost 100,000 km², in some dry seasons the river behaves as an intermittent stream. The monthly minimum maximum ratio is only 0.00155.

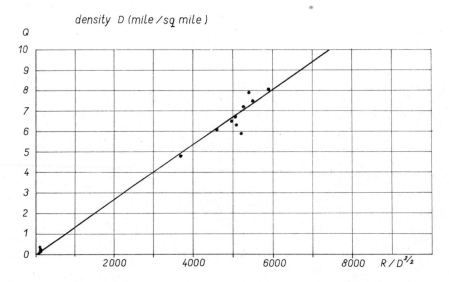

Fig. 4.6. Relationship between the drainage density, mean annual runoff and rainfall pattern, according to Shallash and Starmans.

3. *Dry climate rivers* are under the influence of a pronounced dry season. In these regions, ecologists usually differentiate between wooded steppe and grass steppe, but the river regimes do not differ. They are periodically intermittent and thus the monthly minimum/maximum is 0. In fact, the difference in the regimes between the wet and dry and purely dry conditions can be characterized by an increasing rate of intermittency. Even in areas having more than 1200 mm of annual rainfall, the small streams may behave as intermittent, owing to special morphological and geological conditions. Here, the low density of the drainage network is a good indicator of the corresponding low runoff volume. Shallash and Starmans [24] developed a formula

$$Q = 0.00135 \frac{R^{3/2}}{D}$$

where Q is the mean annual streamflow (cf/mile²), R = total annual rainfall over the catchment (in) and D drainage density (miles/mile²). The formula has been based on observations in the Kafue river basin, Zambia (Fig. 4.6).

4. *Desert climate rivers* are found in the areas with 200 mm and less annual rainfall. Shrubs, desert grass or sand cover the basin surface. The rivers are of the wadí type. The drainage network is very poor and if traceable at all, it was usually developed before the present stage of aridity. Desiccation has resulted in the obliteration of the lower part of the network and in a change from intermittent and perhaps perennial regime to ephemeral. Thus the present stage of the network represents an adjustment to the environment.

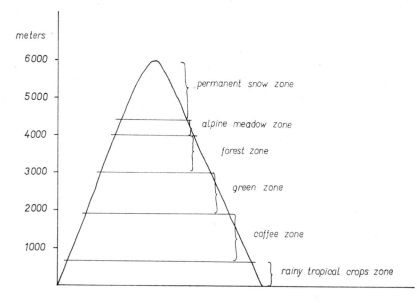

Fig. 4.7. Vegetation of the slopes of tropical mountains.

5. *Rivers of the tropical mountains.* With the exception of Ethiopia, the drainage basin of these rivers is very limited in size. Each of the regimes is a product of very special conditions. The behaviour of the mountain streams is very similar to that of the rivers of the surrounding ecological region, though influenced by the altitude, slope exposure toward the wind and rainfall regime deviations. An example, the Pangani river flowing from the slopes of the Kilimanjaro and Meru (Fig. 4.5) should not be considered representative of all streams of this group. Fig. 4.7 shows the variability of the vegetational cover depending on the altitude of the tropics. River observation in each of the belts might supply very interesting results. The influence of the monsoonal regime can be traced from the shape of the hydrograph in Fig. 4.10. A small difference between maximum and minimum is typical of mountain

streams. A ratio of 0.278 monthly minimum/maximum has been found for the Pangani.

6. *Swamps* are not limited to a certain climatic type and are found anywhere in the tropics. There is no swamp regime which can be considered typical, but vegetational and hydrological regimes are very closely related.

4.3 HYDROLOGICAL BALANCE OF THE MAIN AFRICAN RIVERS

The hydrological regimes of the great African rivers are not part of any of the groups given in the previous section. Many additional factors, mainly changes in the climatic and vegetational belts within the basin, etc., make the regime of each of the great rivers rather unique.

Maximum floods come to the southern tropics in February, March and/or April; north of the equator in July, August and September. A two-peak hydrograph, typical for some rivers in the vicinity of the equator, is gradually transformed into a one-peak hydrograph, but because of special climatic conditions two peaks can sometimes be traced even at a considerable distance from the equator.

The Niger headwaters are near to the ocean in an area enriched by precipitation on the border between the tropical wet and dry zones. The river then flows into the northern semiarid region where it loses water by evaporation in the flats above Timbuktu. From there the river flows into the wet and dry tropical lowland. The increasing delay of the peak can be traced in Tab. 4.1.

Tab. 4.1. Monthly discharges of Niger

	I	II	III	IV	V	VI	VII	VIII	IX	X	XI	XII	Year	Unit
Koulikouro, 1908—52	405	195	105	60	105	390	1230	3255	5235	4320	1890	295	1457	m³/s
Timbuktu, 1924—52	1970	1544	480	447	192	149	213	575	1150	1586	1951	2119	1065	m³/s

In the lowest section, the main peak is delayed by seven months behind the peak of the Upper Niger. A flood is developed downstream under the influence of heavy rainfall in the lower basin which comes at the same time as in the headwaters, but the length of the river and swamps contribute to the formation of a double peak (Fig. 4.8).

The Congo is a river formed by a combination of the tropical highland wet and dry climate and the equatorial wet climate. Tab. 4.2 gives a comparison of the Congo mainstream and its tributaries. The uppermost reach of the Congo, the Chambeshi river above the Bangweulu Swamps, has its peak sometime between March and

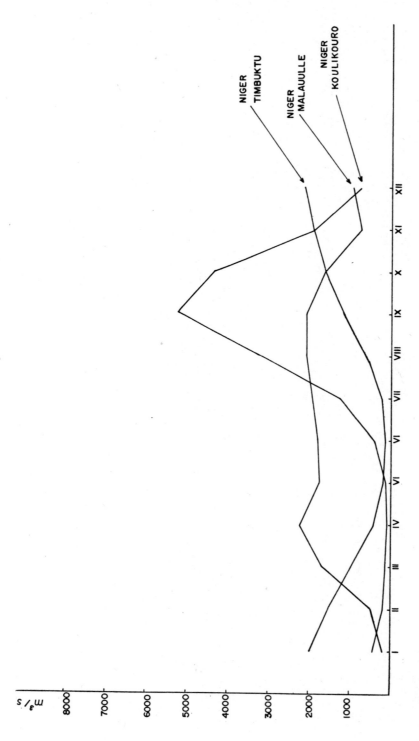

Fig. 4.8. Flood regime of the Niger.

Tab. 4.2. Monthly discharges in Congo basin

	I	II	III	IV	V	VI	VII	VIII	IX	X	XI	XII	Year	Unit
Upper Chambeshi, 1953—65	221	389	591	514	267	155	103	75	53	36	35	83	210	m³/s
Kassai, Mushie, 1932—54	12,900	12,300	12,400	13,200	11,600	8200	6700	6200	6300	7500	10,100	12,600	10,000	m³/s
Ubangi, Bangi, 1910—52	2239	1242	1033	1163	1809	3015	4220	6287	8311	9473	8354	4435	4301	m³/s
Congo, Kinshasa, 1902—14	48.400	38.900	35.200	76.200	39.700	36.800	31.900	32.300	38.500	45.500	54.200	57.500	41.400	m³/s

April. The Kassai, an important tributary in the lower basin under the combined climatic effect found south of the equator, has its first peak already between December and February and its second in March/April. The Ubangi, in many ways a typical equatorial river, has its peak in September. Thus, each of the important tributaries contributes to the formation of the Congo hydrograph at Kinshasa (Fig. 4.9) in its

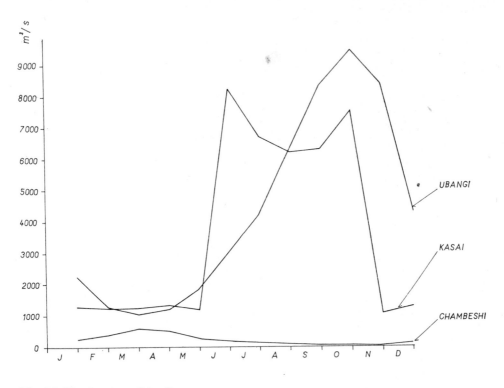

Fig. 4.9. Flood regime of the Congo.

own way. As a result, the main flood comes in November/December and contains the delayed effect of the Congo headwaters. A secondary peak comes in March/April.

The Zambezi is a stream of the main part of the tropical wet and dry highland and semidesert areas of Northern Kalahari. The semidesert area is more than twice the size of the wet and dry highland, however, as the limited rainfall is lost in the desert the contribution from the drainage network is rather insignificant. This results in greatly reduced values of the water balance below the mouth of the Chobe river flowing from the Kalahari area. The March flood peak on the upper Zambezi is delayed by 3−5 months at Lake Kariba, approximately at a distance of 1200 kilometres. A comparison can be seen in Fig. 4.10 and Tab. 4.3.

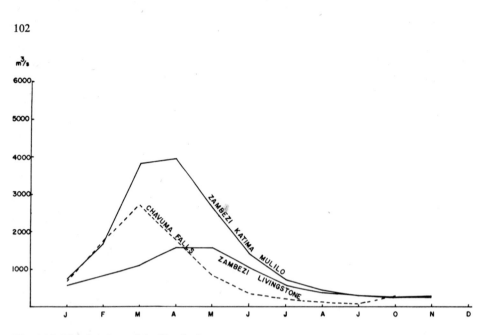

Fig. 4.10. Flood regime of the Zambezi.

Tab. 4.3. Monthly discharges of Zambezi

	I	II	III	IV	V	VI	VII	VIII	IX	X	XI	XII	Year	Unit
Zambezi, Chavuma Falls, 1956—65	687	1757	2727	1818	863	368	219	158	96	73	98	226	755	m³/s
Zambezi, Livingstone, 1908—65	240	244	342	576	840	1633	1600	1585	1025	580	377	278	730	m³/s

Of the main African rivers, *the Nile* has the most complicated regime. The head-waters are located in the so-called undifferentiated highland south and north of the equator. The river flows through tropical wet and dry regions and is transformed into immense swamps. Joined by the Blue Nile, the river flows through semidesert and desert, the effect of which is combined with the artifical reservoirs, irrigation systems etc. Fig. 4.11 indicates how the climatic regions contribute toward the formation of the discharge. Tab. 4.4 gives the monthly discharges for some significant points.

It can be traced from the table how the main lake sources regulate the outflow so that there is no significant annual fluctuation. The influence of the swamps above the confluence with the Blue Nile is also remarkable. They reduce the river discharge by almost one half. Here the seasonal fluctuation becomes more pronounced, owing

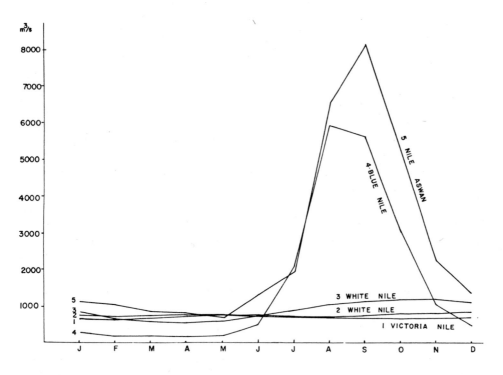

Fig. 4.11. Flood regime of the Nile.

Tab. 4.4. Monthly discharges of Nile

	I	II	III	IV	V	VI	VII	VIII	IX	X	XI	XII	Year	Unit
Victoria Nile, Nama-sagali, 1939—62	628	623	635	670	714	725	693	666	656	644	632	657	662	m³/s
Albert Nile Pakmach, 1955—62	734	696	729	726	711	731	701	691	700	772	791	845	756	m³/s
White Nile, Malakal, 1912—62	829	634	553	525	574	742	897	1030	1130	1200	1200	1200	822	m³/s
Blue Nile, Khartoum, 1912—62	282	188	156	138	182	461	2080	5950	3650	3040	1030	499	1640	m³/s
Nile, Aswan, 1912—62	1100	1020	834	819	698	1340	1910	6570	8180	5200	2270	1400	2650	m³/s

to the increasing influence of the wet and dry climate. Contrary to the November flood on the White Nile, the flood on the Blue Nile comes already in August and is more intensive because of the steep slope of the basin. Observations at Aswan indicate a further delay of the flood peak by one month. The influence of the desert region can be traced from the water balance in Tab. 4.8.

Tables 4.5 — 4.8 give the basic water balance values for the main African streams. The four main streams with a drainage area of more then 10^6 km^2 were balanced. Their basins cover most of the African tropical areas and great variability of the water balance values is found there. The Congo system has the greatest runoff coefficient for the whole basin. It is entirely beyond the influence of the dry climate. On the other hand, the Zambezi and Nile have the smallest coefficients, owing to the influence of the deserts.

Although various reliable sources have been used for the balance calculations, some of the values are still only estimates. More detailed balance values for parts of the main basins can be found in special reports such as those of Bullot [11] and Balek [6, 7].

Since the discharge values for each particular cross section are calculated as long-term means, they may not be found comparable with corresponding values found in various yearbooks, papers, reports etc., as these are based on short periods of observation.

The points of the duration curves for some African rivers are plotted in Fig. 4.12. A dimensionless scale serves as a basis for application of the curves to the unobserved streams for which the mean annual discharge can be estimated by some method. Difficulties usually arise in application of the duration curves owing to the differences in the basin parameters.

4.4 LONG-TERM FLUCTUATION OF AFRICAN RIVERS

Not more than 4% of the land area of many parts of East Africa can expect to receive as much as 1250 mm of rainfall in 4 out of 5 years. About 55% of the land will receive less than 750 mm in 4 out of 5 years and thus the perennial rivers carrying some water all the year are of vital importance to the continent. Unfortunately, no prediction can safely be made as to whether a stream will behave in the next season perennially or intermittently, whether a certain minimum discharge will occur or whether the total runoff volume will be sufficiently high. Long-term forecasting does not go beyond approximate estimates elsewhere in the world either. The discussion on the most suitable models is still far from being closed. Yevjevich [33] suggested simulation of rainfall/runoff sequence by Markov's chains of the 1st and 2nd order. There is scepticism about the existence of relationships between hydrological and hydrometeorological sequences on the one side, and other phenomena such as sunspots on the other [22]. Such a relationship has, however, been indicated experi-

Tab. 4.5. Water balance of Niger basin

River	Location	Dr. area	Preci-pitation	Runoff	Evapo-transpi-ration	Runoff coef.	Water yield	Mean annual discharge
Unit		km^2	mm	mm	mm	%	l/s/km^2	m^3/s
Niger	Sigiri	70,000	1640	420	1220	0.25	13.3	931
Interbasin	Sigiri-Koulikouro	50,000	1424	393	1031	0.28	12.5	624
Niger	Koulikouro	120,000	1550	409	1141	0.26	12.9	1555
Interbasin	Koulikouro-below mouth of Bani	102,600	1235	235	1000	0.19	7.5	770
Niger	below Bani	222,600	1405	328	1077	0.23	10.4	2325
Interbasin	Bani-Benue	501,400	964	35	929	0.04	1.1	532
Niger	above Benue	724,000	1100	126	974	0.11	4.0	2877
Benue	mouth	319,000	1495	343	1152	0.23	10.9	3477
Niger	below Benue	1043,000	1221	192	1029	0.16	6.1	6354
Interbasin	Benue mouth of Niger	48,000	1880	375	1505	0.20	11.9	571
Niger	mouth to Gulf of Benin	1091,000	1250	198	1052	0.16	6.3	6925

Tab. 4.6. Water balance of Congo basin

River	Location	Dr. area	Precipitation	Runoff	Evapo-transpiration	Runoff coef.	Water yield	Mean annual discharge
Unit		km²	mm	mm	mm	%	l/s/km²	m³/s
Chambeshi	above Bangweulu Swamps	43,830	1143	241	902	0.21	7.7	337
Interbasin	Bangweulu Swamps	57,664	1229	59	1170	0.05	1.9	110
Luapula	Below Bangweulu Swamps	101,494	1191	138	1053	0.12	4.4	441
Interbasin	Bangweulu S.-Mweru L.	71,372	1165	136	1029	0.12	4.3	307
Luapula	at Mweru L.	172,866	1181	138	1319	0.12	4.4	754
Kalungwishi	Mweru (mouth to)	26,696	1143	164	929	0.14	5.2	139
Interbasin	Kalungwishi-Lualaba	123,734	1160	132	1028	0.11	4.2	520
Luvua	Confluence with Lualaba	296,600	1172	136	1036	0.12	4.3	1,274
Lualaba	Uzilo	16,300	1100	200	906	0.18	6.3	103
Lufiva	Cornet Falls	11,980	1180	126	1054	0.11	4.0	48
Lualaba	above Luvua	187,800	1130	110	1020	0.10	3.5	651
Lualaba	below Luvua	484,400	1156	126	1030	0.11	4.0	1,931
Interbasin	Luvua-Lukuga	7,200	1125	108	1017	0.10	3.4	24

Lualaba	above Lukuga	491,600	1155	125	1030	0.11	4.0	1,955
Lukuga	mouth to Lualaba	270,900	1062	32	1030	0.03	1.0	271
Lualaba	below Lukuga	762,500	1122	91	1031	0.08	2.9	2,226
Interbasin	Lukuga-Lowani	277,083	1905	602	1303	0.32	19.1	5,295
Lualaba	above Lowani	989,583	1399	239	1160	0.17	7.6	7,521
Lowani	mouth	95,830	1675	274	1401	0.16	8.7	837
Congo	confluence Lowani, Lualaba	1085,413	1422	249	1179	0.17	7.7	8,358
Interbasin	confluence-Ubangi	463,000	1875	482	1393	0.26	15.3	7,126
Congo	above Ubangi	1548,413	1559	315	1244	0.20	10.0	15,484
Ubangi	mouth	754,830	1597	248	1349	0.16	7.9	5,936
Congo	below Ubangi	2303,243	1569	293	1276	0.19	9.3	21,420
Interbasin	Ubangi-Sanga	9,580	1750	355	1395	0.20	11.3	108
Congo	above Sanga	2312,823	1561	293	1268	0.19	9.3	21,528
Sanga	mouth	213,400	1580	362	1218	0.23	11.5	2,471
Interbasin	Sanga-Kwa	109,500	1750	375	1375	0.21	11.9	1,304
Congo	Kwa	2635,723	1581	301	1280	0.19	9.6	3,303
Kwa	mouth	881,887	1538	350	1180	0.23	11.1	9,873
Congo	below Kwa	3517,610	1570	314	1256	0.20	10.0	25,176
Interbasin	Kwa-mouth of Congo	89,840	1300	220	1080	0.17	1.0	629
Congo	mouth	3607,450	1561	313	1246	0.20	9.9	38,805

Tab. 4.7. Water balance of Zambezi basin

River	Location	Dr. area	Precipitation	Runoff	Evapo-transpiration	Runoff coef.	Water yield	Mean annual discharge
Unit		km²	mm	mm	mm	%	l/s/km²	m³/s
Zambezi	Chavuma Falls	75,967	1288	231	1057	0.18	7.3	555
Interbasin	Chavuma Falls-Chobe	284,538	1030	61	969	0.16	1.9	541
Zambezi	above Chobe	360,505	1085	95	990	0.09	3.0	1096
Chobe	mouth	870,758[1]	625[2]	3	622	0.01	0.1	135
Zambezi	below Chobe	1231,263	760	30	730	0.05	1.0	1231
Interbasin	Chobe-Vict. Falls	5,317	605	21	584	0.03	1.1	6
Zambezi	Victoria Falls	1236,580	759	30	729	0.04	1.0	1237
Interbasin	Victoria F.-Kafue	163,380	718	54	664	0.08	1.6	261
Zambezi	above Kafue	1399,960	754	34	720	0.05	1.1	1498
Kafue	mouth to Zambezi	154,856	1023	85	938	0.08	2.7	417
Zambezi	below Kafue	1554,816	782	38	744	0.05	1.2	1915
Interbasin	Kafue-Luangwa	19,091	1198	25	1173	0.02	0.8	151
Zambezi	above mouth Luangwa	1573,907	787	41	746	0.05	1.3	2066
Luangwa	mouth	148,326	925	91	834	0.10	2.9	436
Zambezi	below Luangwa	1722,233	799	44	755	0.06	1.4	2501

[1] with Northern Kalahari
[2] Chobe basin only, 798mm

Tab. 4.8. Water balance of Nile basin

River	Location	Dr. area	Precipitation	Runoff	Evapotranspiration	Runoff coef.	Water yield	Mean annual discharge
Unit		km²	mm	mm	mm	%	l/s/km²	m³/s
Victoria Nile	Ripon Falls	269,000	1302	81	1221	0.06	2.6	699
Semliki	above confluence with Victoria Nile	22,500	1395	88	1307	0.06	2.8	63
Albert Nile	below Albert Lake	281,500	1309	85	1224	0.06	2.7	762
Interbasin	Albert Lake-Mongalla	184,500	1228	20	1208	0.02	0.6	111
White Nile	Mongalla	466,000	1277	60	1217	0.05	1.9	874
Interbasin	Mongalla-Sobat	438,800	900	-38	938	-	-1.2	-511
White Nile	above Sobat	904,800	1094	12	1082	0.01	0.4	362
Sobat	mouth	187,200	1081	71	1010	0.07	2.3	431
White Nile	below Sobat	1092,000	1091	22	1061	0.02	0.7	793
Interbasin	Sobat-Blue Nile	343,000	500	0	500	0.00	0.0	0
White Nile	above Blue Nile	1435,000	710	16	694	0.02	0.5	793
Blue Nile	confluence with W. Nile	324,530	1082	158	924	0.15	5.0	1727
Nile	below confluence with Blue Nile	1759,530	778	43	735	0.06	1.4	2420
Interbasin	confluence-Aswan	79,470	1080	97	983	0.09	3.0	244
Nile	Aswan	1839,000	790	45	745	0.06	1.4	2664
Interbasin	Aswan-mouth	1042,000	7	-18	25	-	-0.6	71
Nile	mouth	2881,000	506	28	479	0.06	0.9	2593

110

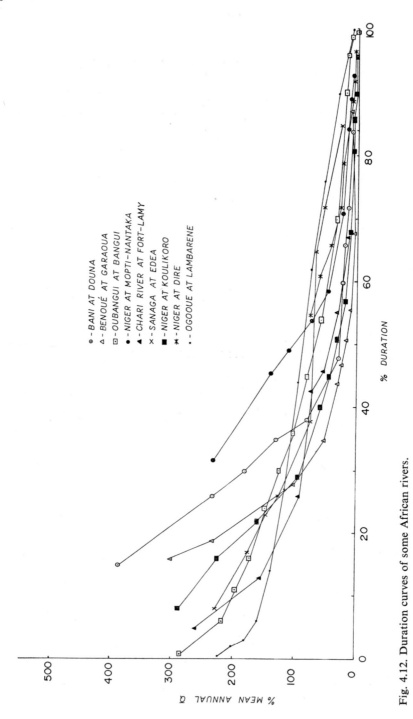

Fig. 4.12. Duration curves of some African rivers.

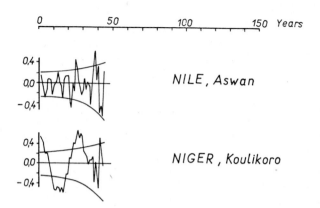

Fig. 4.13. Correlation function for Nile and Niger discharge sequences.

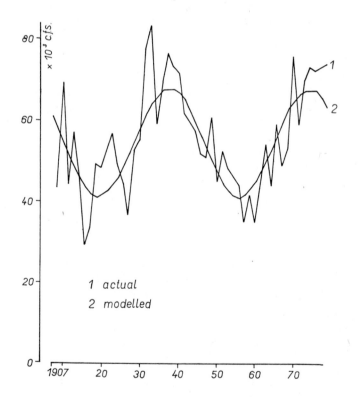

Fig. 4.14. Model of the mean annual discharge sequence of the Niger, based on one main periodicity.

mentally for tropical basins [2]. Unfortunately, very few sequences from the tropics are available for analysis as is evident from the Unesco publications [29] containing available hydrological data on world rivers.

The periodical component of the Niger and Nile sequences was traced by using correlation-function analysis (Fig. 4.13) and Markov's chains of high order. In both sequences, at least ten previous years were found significant for the formation of the mean annual discharge in the eleventh year. For the Niger, the existence of at least one period of 25.5 years was proved at a 95% significance. Periodicities of 7.3 and

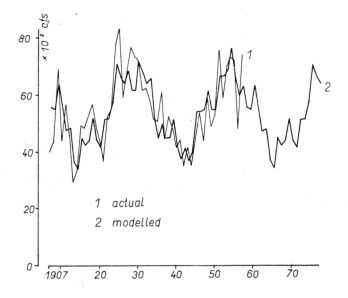

Fig. 4.15. Model of the mean annual discharge sequence of the Niger, based on three main periodicities.

3 years were also traced [1] in the Niger sequence. Periodicities of 21; 7.6; 4.2; and 2.7 years were proved, on the Nile. Taking into account the unequal lengths of the sequences and shapes of the periodograph, a period of 22.6 years has been found significant for both rivers. This period obviously coincides with the period of sunspots categorized by some authors within 22 ± 2 years [17, 32]. Fig. 4.14 gives a model of the mean annual discharge of the Niger as compared with actual observations. When a combination of 25.5, 7.3 and 3 years is applied, the results seem to be even more reliable (Fig. 4.15).

While both analysed sequences are rather small, the longest hydrological record of river stage maxima and minima in existence for the African continent is that for the Nile, recorded by the Arab historians Taghrí Birdi and Al-Hijazi. The sequences are for 849 years with the exception of the Taghrí Birdi sequence of minima, which is for only 663 years. All sequences were adapted and corrected according to changes in the cubit and finger scales within the period of observation, differences between the Christian and Mohammedan calendar and fluctuations of the river bed. There may still be some errors in the sequences, but even so, they can be considered unique material for both hydrologists and Egyptologists [3]. Periodicities traced in the sequences are given in Tab. 4.9.

Tab. 4.9. Periodicity in Nile historical sequences

Author	Taghrí Birdi		Al Hijazi		Hurst
Sequence of unit	minima	maxima	minima	maxima	runoff volume
	m	m	m	m	mld m^3
Observed at	Roda	Roda	Roda	Roda	Aswan
Length	663	849	849	849	84
Periodicity	440	556	556	556	20.5—32.2
traced	189—265	242—389	242—389	242—389	7.3
(interval of years)	6.6	151—332	89—113	132—151	
		14—14.2	74—82		
			16.2—18.8		

Longer periodicities as given in the table may reflect such phenomena as rising of the river bed or the existence of even longer periodicities. It is curious that the number seven plays a particularly significant role in the short periods. The periodicities 7.3 years and 21.7 years of the interval 20.5—32.2 years, traced in Hurst's short but more accurate sequence, can be related to the periodicities of 77 years (interval 74—82 years) as found in Al-Hijazi data and to the periodicities of 6.6 and 14.1 (interval 14—14.2 years) in the Taghrí Birdi data. The periodicity of 18.4 years (16.2—18.8 years interval) may be influenced by several doubled values as given by Al-Hijazi for a leap-year difference between the solar and lunar calendar once in 33 years.

Plotted in Fig. 4.16 are the sequence of both historical records based on the corrected values of Taghrí Birdi. The autoregressive analysis was used to find the number of year-lags significant for the discharge in the year which follows. This is given in Tab. 4.10.

An autoregressive model based on the Taghrí Birdi minima sequence can be written as

$$y_t = 0.42y_{t-1} + 0.10y_{t-2} + 0.06y_{t-3} + 0.10y_{t-5} + 0.02y_{t-6} + \\ + 0.04y_{t-11} + 0.11y_{t-13} + \varepsilon_t$$

114

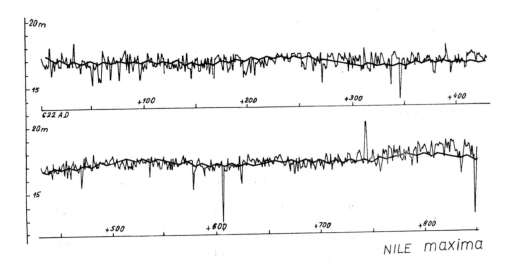

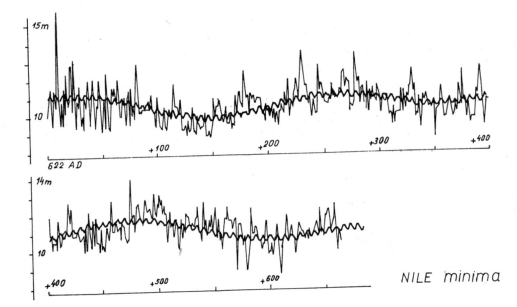

Fig. 4.16. Sequence of Nile maxima and minima according to Taghrí-Birdi.

Tab. 4.10. Lags significant for the autoregressive models of Nile maxima and minima sequences

Author	Taghrí Birdi		Al Hijazi	
Sequence of	maxima	minima	maxima	minima
Lag significant	1	1	1	1
	2	2	2	2
	3	3	3	3
	4	5	4	4
	6	6	5	5
	7	7	6	6
	8	11	7	7
	9	13	8	
	10		9	
	13		15	
	17			

where $y = x - 10.8$ (m) is the deviation from the mean value. Similarly, an autoregressive model of the sequence of maxima is

$$y_t = 0.20y_{t-1} + 0.07y_{t-2} + 0.09y_{t-3} + 0.06y_{t-4} + 0.04y_{t-6} + 0.02y_{t-7} + 0.05y_{t-8} + 0.08y_{t-9} + 0.06y_{t-10} + 0.10y_{t-13} + 0.09y_{t-17} + \varepsilon_t$$

where $y = x - 16.99$ (m) is the deviation from the mean value. The metric system has been used.

Apart from the first terms of both equations, the lags of 13 are obviously more significant than the others. Together with the observed values, 14 years is again obtained in both cases.

Discussing the long-term fluctuation of the hydrological and hydrometeorological sequences in general, it must be said that the various sequences observed in the wet and dry climatic regions of the southern African hemisphere do not indicate a direct relationship between rainfall and runoff. For instance, in the sequence of the mean annual discharges of the Kafue river [5] a periodicity of 7.6 years was proved, while in the rainfall records from the Kafue basin, periodicities of 3.4, 4.4 and 5.4 years were found.

It can be concluded from the results that certain periodicities appear to be more pronounced close to the equator, while with increasing distance from the equator a combination of more or less equally significant sequences is found. Long-term fluctuation of sunspots and water levels and their mutual relationship has long been the subject of study. Brooks [10] first pointed out that such a relationship may exist and this was later confirmed by Gregory [13]. On the other hand, Hurst, a leading Nile hydrologist, did not credit this.

Kuzin [18] analysed the streams of North Africa and classified the rivers according to their seasonal fluctuation into three zones as seen in Fig. 4.17.

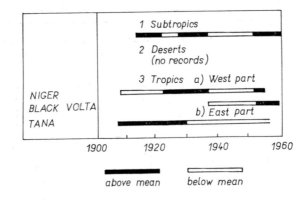

Fig. 4.17. Kuzin's analysis of long-term fluctuation of African rivers.

Anděl and Balek developed a graph of the disturbances observed in the most significant periodicities of some world rivers. The results for the Nile and Niger are given in Fig. 4.18.

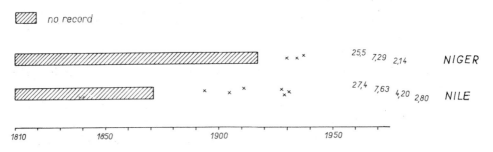

Fig. 4.18. Disturbances in the Nile and Niger sequences.

4.5 SOIL EROSION IN THE AFRICAN BASINS

No systematic study on the soil erosion of the African continent has been reported so far. The most systematic measurements have perhaps been made in the Nile basin. Simaika [25] presented a graph of a typical Nile flood and the fluctuation of silt movement (Fig. 4.19). Since the Nile is a stream with very different regimes of dissolved solids, the values obtained there must be used with caution outside the basin.

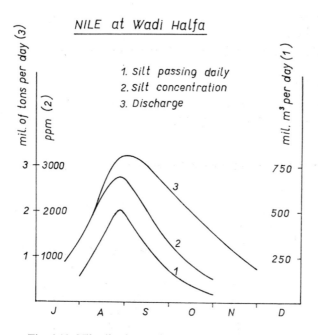

Fig. 4.19. Nile silt observation according to Simaika.

First data on the dissolved solids were collected by Hurst for various profiles in the Nile basin (Tab. 4.11):

Tab. 4.11. Dissolved solids in various cross sections of Nile basin

Lake Victoria	80 ppm (by weight)
Victora Nile below Kyoga	100 ppm (by weight)
L. Edward	670 ppm (by weight)
L. Albert	590 ppm (by weight)
Lake Tana	170 ppm (by weight)
Albert Nile below Albert L.	160 ppm (by weight)
Blue Nile at Khartoum	140 ppm (by weight)
White Nile at Kartoum	130 ppm (by weight)
Atbara	200 ppm (by weight)
Nile at Cairo	170 ppm (by weight)

Starmans [26] analysed the erosional processes in several African basins. He regarded the following factors to be decisive in tropical erosional processes:

Aridity and/or semiaridity of climate,

degree of destruction of vegetation,

ability of vegetation to regenerate,

leaf-width of vegetation,

state of soil-protection against violent temperature changes,

age of rivers,

regime of rainfall.

The following silt loads were observed by Starmans in the Malawian streams during the period 1953−1954 (Tab. 4.12):

Tab. 4.12. Silt loads of some Malawian rivers

River	Silt collecting area miles^2	Silt loads tons	Silt tons per m^2
		measured from January 1, 1953 to June 30, 1954	
Lirangwe	78	23,876	306
Lunzu	38	13,767	362
Kwakwasi	25	4,012	160
Tuchile	240	254,245	470
Nswadzi	48	20,129	419
Mwaphanzi	117	29,650	253
Likabula	218	44,001	202
Mudi	7.25	13,082	1804
Maperera	25	9,058	362
Mwanza	630	8,630	14
Wankulumadzi	205	58,044	283
Lisungwe	461	78,833	171
Rivi Rivi	377	60,340	140
Sombani	289	630	2
Mulungushi	9.2	1,932	214
Palombe	537	880,176	1639

During the same period, a measurement of dissolved solids was made with the following results (Tab. 4.13):

The relationship between the percentage of catchment adequately covered by vegetation and silt load was developed by Starmans and is given in Fig. 4.20. He concluded that although the hilly areas of Africa produce quite high floods of short duration, the effect of the vegetational cover is evident. It is worth noting that the amount of dissolved solids exceeds the amount of suspended solids in some basins.

The positive influence of vegetation on catchment stability is also indicated by the results obtained by Balek and Perry [8] on four heavily afforested catchments

Tab. 4.13. Dissolved solids of some Malawian rivers

River	Total tonnage	Tons per sqm
Lirangwe	10,407	135.5
Lunzu	6,439	169.4
Kwakwasi	5,736	190.9
Tuchile	67,278	124.5
Nawadzi	26,840	555.7
Mwaphanzi	34,769	296.4
Likabule	32,381	148.2
Mudi	1,347	185.0
Maperera	16,083	653.7
Mwanza	51,745	82.1
Wankulumadzi	42,500	207.2
Lisungwe	44,322	96.0
Rivi Rivi	24,471	64.8
Sombani	7,144	24.7
Mulungushi	2,026	220.2
Palombe	88,457	164.8

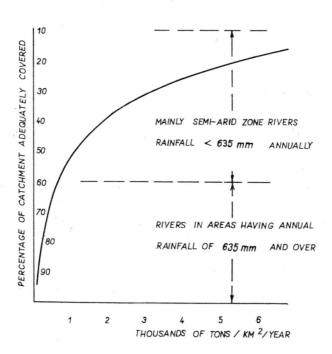

Fig. 4.20. Relationship of vegetation and soil erosion in selected rivers of Africa and Asia.

in Zambia. During the 1970/71 rainy season, the amount of soil washed from the areas covered up to 90−95% by tropical highland forest was between 0.000053 − 0.000142 mm only.

The load carried by the West African rivers was measured by Grove [14] during the Trans-African Hovercraft Expedition. Measurements in the rivers Senegal, Niger, Benue and Shari were based on a chemical analysis. As the river regimes had such

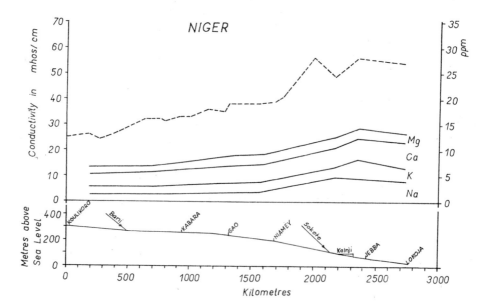

Fig. 4.21. Dissolved load of Niger, according to Grove.

complications as loss of water downstream, evaporation from the water surface etc., results were considered by the author as only indicative. Grove's estimate for the Senegal flood period was 40 ppm of dissolved solids, while for the Benue this amount was estimated as the mean. Fig. 4.21 plots the fluctuation of the dissolved load in the Niger.

Fig. 4.22 gives the relationship between the discharge and the dissolved load as measured by Roche in the Shari river [21]. The results indicate that the concentration is at its lowest point at the beginning of the rainy season in July and rises to its peak value the next May. On an average, 2 million tons of dissolved solids are carried into Lake Chad every year. A similar amount from other sources, such as dust, is carried into Lake Chad year after year.

Gallois [12] puts the estimate at 2.4 million tons of suspended load per year carried by the Niger at Koulikoro and 1.2 million tons of dissolved load at the same point.

An interesting comparison is given by Grove for some West African rivers (Tab. 4.14). A much lower erosion rate was observed in Malagasy by Touchebeuf [27].

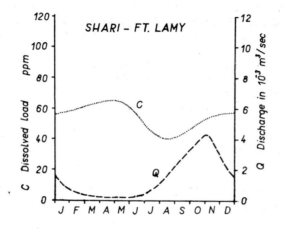

Fig. 4.22. Discharge and dissolved load of Shari river, according to Roche.

Tab. 4.14. Erosion rates for some African rivers

River	Niger	Benue	Niger-Benue	Congo
Discharge 10^{-12} m^3/year	0.1	0.11	0.21	1.2
Erosional rates 10^{-6} tons/year				
in solution	5.5	4.5	10.0	98.5
suspended	9	22	31	31.2
combined	14.5	26.5	41	129.7
tons/year/km^2	19	77	37	37
10^3 mm/year	8	36	15	14.8

4.6 LIST OF LITERATURE

[1] Anděl, J., Balek, J., 1970. Mathematical-statistical analysis of the formation of the hydrological sequences (In Czech), *Vodohospodársky časopis,* Bratislava pp. 3—28.
[2] Anděl, J., Balek, J., 1971. Analysis of periodicity in hydrological sequences. *Journal of Hydrology* 14, 1971, 66—82.
[3] Anděl, J., Balek, J., Verner, M., 1971. An analysis of historical sequences of the Nile maxima and minima. Symp. on the role of hydrol. in the econ. dev. of Africa, Addis Ababa, 15 p.
[4] Balek, J., 1968. Linear extrapolation of the mean annual discharge of rivers. (In Czech). *Vodohospodársky časopis* XIV, 3, 1968, Bratislava, 10 p.
[5] Balek, J., 1970. An analysis of the hydrometeorological sequences observed in the Kafue River Basin. NCSR Lusaka TR5/WR5, 23 p.

122

[6] Balek, J., 1970. Water balance of Luapula and Lake Tanganyika basin. NCSR Lusaka TR Rep., 24 p.

[7] Balek, J., 1971. Water balance of Zambezi basin. NCSR Lusaka TR Rep., 35 p.

[8] Balek, J., Perry, J., 1972. Luano catchments, first phase final report. NCSR Lusaka, TR, p.

[9] Bliss, E. W., 1925. Nile Flood. World Weather Mem. of Roy. Met. Soc., Vol. 1, No. 5.

[10] Brooks, C. E. P., 1923. Variation in levels of the Central African Lakes. Geophys. Mem., No. 20.

[11] Bullot, F., 1970. Climatique du Basin Congolés. Inst. Nat. pour l'étude agron. du Congo.

[12] Gallois, J., 1962. Le delta intérieur du Niger. Cent. de Rech. et Doc. Cartograph et Géog. CNRS, Paris, Vol. 3, 153 p.

[13] Gregory, R., 1930. Weather Recurrence and Weather Cycles. *Quat. Jour. Roy. Met. Soc.,* 1930.

[14] Grove, A. T., 1972. The dissolved and solid load carried by some West Afr. rivers. *Journal of Hydrology* **16**, 1972, pp. 287—300.

[15] Keller, R., 1962. Gewässer und Wasserhaushalt des Festlandes. Teubner Verlagsgesellschaft, Leipzig, 519 p.

[16] Kimble, G. T., 1960. Tropical Africa. The Twentieth Cent. Fund, N. York 1960.

[17] Kopecky, J., 1967. The periodicity of sunspot groups. Adv. in Astr. and Astroph., Vol. 5, N. York.

[18] Kuzin, N., 1970. Cyclic variations of river runoff of the Northern Hemisphere (In Russian). Gidrometizdat, Leningrad.

[19] Lvovich, M. I., 1945. Elements of the water regime of the earth. Moscow (In Russian).

[20] Popper, W., 1957. The Cairo Nilometer. Univ. of Cal. Public, in Sem. Phil, vol. 12, Berkeley.

[21] Roche, M. A., 1969. Evolution dans l'espace et le temps de la conductivité électrique des eaux du Lac Tchad. Cah. Orstom, sér. hyd., 43. p.

[22] Rodriguez, I., Yevdjevich, V., 1968. The investigation of relationship between hydrological time series and sunspot numbers. Col. State Univ., Hyd. Pap. 26, 1968.

[23] Shahin, M., 1971. Hydrology of the Nile basin. Int. Courses Delft, 140 p.

[24] Shalash, S., Starmans, G. A. N., 1969. Estimation of mean annual streamflow from precipitation and drainage density. Lusaka, Water Affairs Dept., 9 p.

[25] Simaika, Y. M., 1940. The suspended matter in the Nile. Cairo, Schindler's Press.

[26] Starmans, G. A. N., 1970. Soil erosion of selected African and Asian catchments. Int. Water Er. Symp., Prague, 10 p.

[27] Touchebeuf, D., 1961. Étude de transport solide en Afrique et á Madagascar. Symp. on Af. Hydrol. Nairobi, 12 p.

[28] Tousson, O., 1925. Memoiré sur l'historie du Nil, t. 3, MIE, VIII, 1925, pp. 265—266.

[29] UNESCO 1972. Dicsharge of selected rivers of the world, Paris, 3 parts, 362 p.

[30] Voeykov, A., 1884. Flüsse und Landseen als Produkte des Klimas. Die Klimate des Erdballes. St. Petersburg.

[31] Whyte, R. O., 1966. The use of arid and semiarid lands. UNESCO Symp. on arid lands.

[32] Willet, A., 1960. Long term indices of solar activity. Sc. Rep. No. 1, NSF, Cambridge, Mass.

[33] Yevjevich, V. M., 1964/64. Fluctuation of wet and dry years. Two parts, Colorado State Univ., 105 p.

5. GROUNDWATER RESOURCES

5.1 GENERAL

In special conditions of arid, semiarid and wet and dry regions, groundwater is one of the most important natural resources. In humid and extra-humid parts of the tropics groundwater also plays an important role since many surface sources are intermittent or polluted and infected and thus local supply needs, cattle watering, irrigation and industry depend on groundwater sources. Surface water is also not good for consumption in areas with abundant rainfall because of the presence of various bacteria and parasites and of contamination. The developing industry in the rapidly growing cities requires more water than was foreseen at the time of the towns' establishment. Since more than fifty percent of African cities depend on groundwater supply, the problem of groundwater depletion is becoming rapidly more serious even in the tropics.

The survey of groundwater resources in tropical Africa can be characterized as lacking uniformity. The tropical parts of South Africa, for instance, have been extensively and systematically surveyed. Much attention has also been paid to the arid regions, particularly the parts of Sahara under Arab influence. French teams have been using advanced modern techniques [6], including isotopes, while prospecting in the regions once under British rule has been carried on with simple geophysical instruments. The countries in French-speaking Africa are more arid and the groundwater problem thus more vital.

According to the UN report [22] on the tropical African countries, less than 360 sets of drilling equipment were available and only 50 specialized groundwater engineers were making surveys there. Certainly these figures have increased since then, but there is still a shortage of qualified staff and instruments.

The statement by Bisset [5] can serve as an example of groundwater requirements in tropical humid regions, based on long experience in Uganda: "However impressive the large bodies of water may appear on the map, they have little direct effect beyond the distance from their banks over which the water can be carried on the heads of women. The bulk of the population relies for its supplies not on these large lakes and rivers but on the local water hole."

5.2 GROUNDWATER STORAGE, ITS DEPLETION AND REPLENISHMENT

The possibilities of groundwater-recharge in the tropics are very limited. There are three possible sources apart from the water stored during the past geological periods:

a) condensation of atmospheric humidity,

b) rainfall,

c) infiltration of surface water from streams and lakes.

The recharge by condensation is considered insignificant in the tropics. The recharge from rainfall is significant where the rainfall coincides with increased humidity, otherwise any recharge is reduced by evaporation. Rock formation also plays an important role. Denuded bare rock containing fissures increases the recharge significantly, whereas on the contrary, vegetation with a large leaf area reduces it either directly by evaporating the intercepted water, or indirectly by transpiring the stored groundwater. At this point, the negative role of phreatophytic plants should be recalled; they tap the water for no or little economic value and waste it. Almost any tropical vegetation plays an important role in groundwater depletion.

Replenishment of groundwater by intermittent streams may exceed the contribution of rainfall to the rivers with sandy or graval beds or fissures. The drinking-water supply of the Gwembe Valley, an area north of Kariba reservoir, depends for instance, on the groundwater accumulated in the river beds. The seepage from surface runoff flowing through temporary river channels contributes in some desert areas up to ten times more than the rain itself. As stated by Dubief [10]: "Only from that proportion of rainfall evacuated as runoff with consequent concentration in limited zones in appreciable depth and for a fair space of time can there be a deep infiltration which will replenish the desert water table."

Tisserout [24] differentiates in arid tropics between two sorts of replenishment, depending on the type of surface runoff:

a) saturation runoff, resulting from saturation of the soil after prolonged rainfall,

b) intensity runoff, when the rainfall is too intensive for the infiltration process.

In some years with heavy rainfall, the aquifer may be depleted more than in periods with less rainfall but more beneficial to the aquifer.

Martin [18], studying the groundwater in the Kalahari sands, credits all replenishment to the account of local percolation. The best arid areas available for replenishment are considered to be those with a uniform water table between 6–30 meters, a sand cover not exceeding 0.3–1.2 meters and the layer of calcrete overlying the Kalahari beds. It has been proved in the Northern Gobabis District, east of Windhoek, that when the sand layer rises to over 10 meters, the groundwater level shows very irregular behavior.

In the small pore spaces of sand, about 500 mm of rainfall can be easily infiltrated and than the evapotranspiration consumes the whole of the water available.

Fig. 5.1 indicates the configuration of the groundwater table above the Kalahari beds between the rivers Otjozondjou and Epukiro. The water table slopes quite steeply from the plateau toward the valleys, nevertheless, there are no springs in the valleys. According to Martin, this proves a static equilibrium between permeability and the hydrostatic head or, in other words, the lowest permeability at the steepest areas. A similar phenomenon has been observed in East Africa. The generally accepted idea that with humid conditions, such as those existing in parts of Uganda, the groundwater should be found as a continuous body the top of which follows the contours of the land and comes nearer to the surface of the ground in the valleys, has been proved not to be always valid. Instead, the water has been found at greater depths below rises than in the valleys and in pockets of variable extent, type and depth.

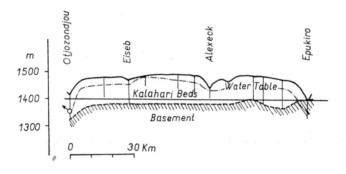

Fig. 5.1. Position of water table in Kalahari Beds between Otjozondjou and Epukiro. Vertical lines mark the boreholes. After Martin.

While the process of evaporation from the soil is more or less known, there have been far fewer experiments to determine evaporation from the groundwater table in deserts, through the various sand layers. Helwig [13] proved that a drop of the groundwater table below 60 cm in sand with a mean diameter of 0.52 mm will practically prevent evaporation losses. In the area of the Kalahari desert, he measured evaporation from the free surface, which was higher by 8% than that from saturated sand. A series of results (Fig. 5.2) indicates a dependence of daily evaporation on the depth of the groundwater table. Daily fluctuation of evaporation indicates a high degree of variability even for a constant groundwater level (Fig. 5.3).

A frequent source of water supply in arid and semiarid regions are the beds of dry rivers, which can be characterized as shallow regions in the model discussed in Chap. 3. The special structure of the river bed makes the evaporation/evapotranspiration process different from those in deep sand layers. The concentration of phreatophytes is one of the factors playing an important role. In a series of experiments on the

Swakop intermittent river in Kalahari, Hellwig proved that on an average 67.9% of water loss is accounted for by evapotranspiration, while 19.5% evaporates from permanently wet barren sand and 12.6% from temporarily wet sand. A substantial

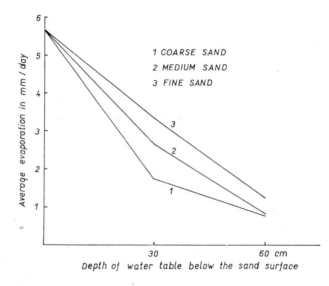

Fig. 5.2. Evaporation from sand as a function of the depth of the water table. After Hellwig.

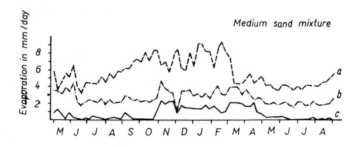

Fig. 5.3. Fluctuation of the evaporation from sand as a function of the depth of the water table in medium sand mixture. After Hellwig. (a) water table at the sand surface, (b) 30 cm below sand surface, (c) 60 cm below sand surface. After Hellwig.

part of evaporation/transpiration is due to well-adapted phreatophytes with their roots in the permanently wet sand. The results, however interesting, cannot be generalized without additional measurements being made not far from the studied regions.

5.3 GROUNDWATER IN THE DESERT

According to Dixey [8], the aridity and humidity of an area depends not only upon the climatic belts within which the area is located, but also upon the climatic belts in the surrounding region and upon geological and geomorphological events and when these are of a late geological age, it is clear that they can be of considerable importance in the present context with the existence of a prevailing desert regime such as that of North Africa; it is quite possible for geological and geomorphological processes to alter profoundly the climate of the parts of the region which would be reflected in a lack of harmony between the observed features of the soils, water supply etc. and the current climate. The lowlands of Northern Kenya form a belt of arid plains extending from Somalia and Ethiopia to the Sudan and the Sahara, which has been arid since at least Tertiary times. Gentle downwarping created Lake Rudolph. Geomorphological uplifts in many parts of Africa may have resulted in increased precipitation and a decrease of aridity.

Both the quality and the quantity of groundwater has been influenced by the geological processes. It is accepted that an uplift of large parts of the continent in the Quaternary epoch reduced the salt content of waters while the profile of the depressed land masses was impregnated with salt and developed groundwater salinity.

Artesian waters have a special importance in the arid tropics, though their utilisation has not yet been fully developed. Either natural or drilled artesian water outflows, forming lakes and swamps, are known as shotts in Algeria and shebkas in Libya. In Senegal too, artesian water is one of the country's important natural resources. In this respect it should be pointed out that the African desert areas, with a precipitation between 0 and 250 mm, have one of the world's greatest aquifers. As estimated by Nace [19] in the Nubian sandstone and Continental Intercalary Fountain, on an area of 6.5×10^6 km^2, about 600×10^3 km^3 of water are stored. The current recharge of these reservoirs is negligible and they were formed during the pluvial period at the end of Pleistocene, which means that the age of the water is 30,000 to 40,000 years. The sources have been occasionaly tapped by hand by the local inhabitants down to depths of almost 150 m, while in the arid parts of South Africa the water has been traditionally collected by so-called air wells, as it condensed on the surface of limestone blocks.

5.4 GROUNDWATER IN WET AND DRY REGIONS

Though available groundwater resources are in general more rich than in the arid regions, there are occasionally problems as to the availability of the water, particularly at the end of the dry season or during the year following a dry year in which the rainfall total was insufficient to replenish the groundwater storage. Even wells constructed in accordance with long-term experience may then dry out. In some

128

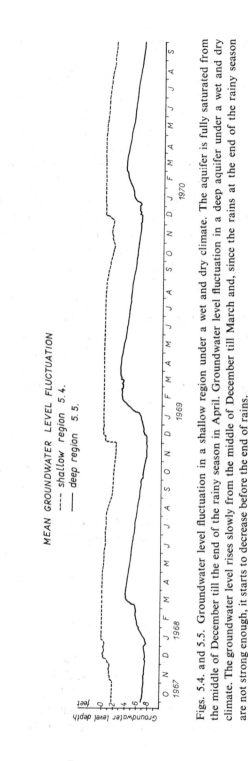

Figs. 5.4. and 5.5. Groundwater level fluctuation in a shallow region under a wet and dry climate. The aquifer is fully saturated from the middle of December till the end of the rainy season in April. Groundwater level fluctuation in a deep aquifer under a wet and dry climate. The groundwater level fluctuation in a deep aquifer under a wet and dry climate. The groundwater level rises slowly from the middle of December till March and, since the rains at the end of the rainy season are not strong enough, it starts to decrease before the end of rains.

areas with an otherwise reasonable groundwater storage, drilled boreholes may fail to supply water due to the unfavourable geological structure of the aquifers. On the left bank of Kariba, for instance, half of the approximate number of 200 boreholes and wells failed to produce an adequate amount of water and some of them had no water at all. In this and other areas, sociological aspects come into play since the inhabitants who traditionally use river water, refuse to drink water from the wells, calling it dead, and then prefer to wait patiently near the sandhole in a dry river bed for it to fill.

In the wet and dry regions, two groundwater regimes can usually be recognized according to the soil profile and the type of aquifer. Some wells indicate a higher degree of sensitivity toward the precipitation regime and the groundwater level in these areas rises rapidly quite soon after the rainy season starts. After the aquifer becomes fully saturated, the groundwater level remains almost constant until the end of the rainy season and in some cases a few days longer (Fig. 5.4). Groundwater in the shallow regions formed by the river beds, intermittent swamps and sandy aquifers with impervious layers in the vicinity of the surface, behave according to this pattern.

In the woodlands and in areas with deeply situated aquifers, etc., the groundwater starts to rise several weeks or months after the onset of the rainy season and it continues to rise for several weeks or months, as in the case of the deepest crest sites. Sometimes the maximum level is not reached until the end of the rainy season (Fig. 5.5), but this depends on the distribution of the rainfall and intensity at the end of the wet season and on the evapotranspiration regimes and their relationship toward the accumulated groundwater storage. When there are several observation wells in a relatively small region, one can find that the groundwater regimes, particularly the commencement of the rise and the peak, are not identical even in a small area which is seemingly orographically uniform. Usually, tropical vegetation, its distribution, type of species etc. is responsible for this type of variability. A joint evaluation of the level fluctuation of all the wells in the area, by means of Thiessen's polygon or an equivalent method, is recommended in such cases.

5.5 AFRICAN WATER-BEARING UNITS

As follows from the previous chapter, groundwater occurrence depends on the past and present climate; the geological structure of the continent, with lowlands in the North and West and highlands in the East and South, contributes to the uniformity of the groundwater regimes on the continent. The climatic characteristics usually vary only between extremely humid, wet and dry and arid regions. The geological structure is more complicated. The greater part of the continent consists of a Precambrian crystalline and metamorphic shield and three main units are recognized by geologists: Lower Precambrian, prevailingly consisting of granitoids and granite gneiss;

Middle Precambrian, essentially schist-quartzitic and eruptive; and Upper Precambrian with schists, sandstones, lavas and conglomerates. The outcroppings of the basement correspond to the uplift areas which encircle the depressed basins with sedimentary formations, such as the Niger, Chad, Nile, Congo, Kalahari. Within these areas, Infracambrian and Cambrian formations of marine origin, composed mainly of calcareo-dolomitic rocks and calcareous shales, found mainly south of the Equator, particularly at Katanga and metamorphosed in Zambia, have been identified at the lower levels. The younger formations consist mainly of shales and sandstones with some limestone and clay beds. The Somalian plateau and some coastal sedimentary basins were covered by the Jurassic and Lower-Cretaceous seas. During this period, called the Continental Interclaire, formations mainly of sandstone were deposited in most of the basins. During the Tertiary period, formation of sandstones and sands developed. Finally eruptions occurred on the breaking lines during the Miocene and Pliocene along the Rift Valleys, and the North Atlas was accreted to the continent during the Alpine orogeny.

In general, large water-bearing units correspond to large geological bodies. Precambrian formations of large peneplains, at low altitudes in West Africa and higher altitudes on the East African highlands, have their water bearing aquifers in weathered, fractured or combined sections. The upper horizon is frequently formed by laterites and argilosands temporarily containing groundwater. The next horizon is formed by kaolinic porridge. Below are the weathered sections of bedrock holding the water under a certain head. The effectiveness of wells is therefore related to the locality chosen. While the schists are usually sterile, the quartzites, dolerites and vein-rocks have better yields.

There are important water-bearing units in the Infracambrian and Paleozoic formations. The aquifers of the calcero-dolomitic and limestone-shale series are most significant, being found south of the Equator in Katanga and parts of West Africa. In the Lusaka area in Zambia, more than 1000 m^3/hour is pumped in seasonal peaks from these aquifers and the yield of a single borehole is about 100 m^3/hour. Here sheet cavitation exists at some depths, according to Lambert [17]. The cavitation is believed to be due to the water table falling during various interpluvials in the Pleistocene. Even when of small dimensions, the enlarged cavities of the fissures are

The large sedimentary basins of Central Africa are filled with soft deposits of Paleozoic, Jurassic and Cretaceous origin. Some of them, e. g., in the Chad region, and Northern Nigeria, contain strong sources of artesian waters. The Kalahari sandstones may also contain many aquifers. In the Sahara region (Fig. 5.6) intermediate and deep aquifers are found in the continental sandstones. Important artesian wells have been established in these regions. The water is usually termed very significant. The aquifers of the Cambroordovician shales, sandstones and schists also bear water in fractures; usually the more fractured schists have a better yield. fossil, although some possibility of replenishment from natural sources or from the backwater of the Aswan reservoir may exist. In Sudan this water is found in less-extensive sandstone localities. The Saharan Cretaceous limestone aquifers and alluvial aquifers which are replenished occasionally by flooded wadis, are of a small size. However, most of the water was stored there in the past millenium, during the pluvial periods when the Sahara region experienced far more rain than today. The

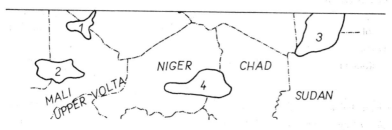

Fig. 5.6. Distribution of the groundwater reservoirs in tropical Sahara. The storage capacities are estimated for 1. Tanezrouft — 0.4×10^3 km^3, 2. Niger — $1.8. \times 10^3$ km^3, 3. Western Egyptian Desert — 6×10^3 km^3 (including subtropical part), 4. Chad — $3.5. \times 10^3$ km^3. More reservoirs exist in North Sahara and other reservoirs can also be located.

sedimentary coastal basins spread over Mauretania, Senegal and the Ivory Coast basin, Ghana, Togo, Dahomey, Nigeria and parts of Cameroon, Gabon, Congo, the Angola basins, Mozambique and the East Africa region, including also part of Somalia and part of Madagascar. Sandstones, shaly limestones, sands and alluvial deposits in many cases give a high yield, however near the coast the water may be mixed with sea water.

Vast formations of the Mesozoic plateau of Ethiopia-Somalia, the lava fields of Ethiopia and the Rift valleys of East Africa are still to be explored, except for some small localities. The water yield in these regions is very variable. Nevertheless, a great number of hydrogeological explorations has been accomplished in Africa. Here at least some of the projects carried out by the United Nations Agencies in past years are listed: Study of groundwater resources in the Northern Sahara, groundwater prospection and pilot development in Cameroon, pilot groundwater development in Dahomey, pilot project in groundwater utilisation in New Valley area in Egypt, groundwater study in Nairobi, survey in mineral and groundwater resources in Southern Malgasy, study of a sandy Maestrichtian aquifer and coastal terminal in Senegal, mineral and groundwater survey for Somalia, groundwater survey in the coastal region of Togo.

Numerous studies have been accomplished on a national scale and some of the most significant are listed in the excellent bibliography of Rodier [1].

The Resources and Transport Division of the United Nations prepared a table (Tab. 5.1) as a guide giving basic information on the water yield in some geological formations of Africa.

Tab. 5.1. Estimated yields of selected African aquifers

Formation	Average yield m³/hour	Maximum yield m³/hour
Precambrian backbones		
Granitogneiss	2—5	20 and more
Schists, rhyolites, diorites	less than 1	no information
Infracambrian and Paleozoic		
Shale-limestone and calcalcareo dolomite	10—100 (depending on degree of fracturation)	no information
West African shales and shale-limestones		no information
Large sedimentary basins of Central Africa		
Karoo sandstones	1	200
Artesian waters of N. Nigeria		1000
Kalahari sands	1—10	up to 50
Cenezoic and quaternary deposits of Congo and Chad basins	1 to small	no information
Sahara		
Continental Interclaire	no information	100 and more
Alluvium	small	up to 1000 (Nile region)
Dunar sands	less than low	
Sedimentary coastal basins		
Senegal	20	100
Ivory coast	no information	200
Togo	20.	
Somalia	no information	10 to small
Madagascar	no information	up to 300

5.6 THE PROSPECTION AND UTILIZATION OF GROUNDWATER RESOURCES

Economic utilization requires a detailed qualitative and quantitative survey with the aim of obtaining a reliable estimate. Drilling and observation of the wells and boreholes and pumping tests are the most common methods. There are numerous techniques for selecting the sites for drilling and they are described in the specialist literature. The survey must supply an answer to three basic questions:

a) where the water is,

b) how much is available,

c) how much can be withdrawn before exhausting the aquifer.

Until recently, empirical methods have been most often used in Africa and most drospectors searched for the groundwater supply without considering the groundwater regime of the whole region. At present, various methods applied elsewhere in Africa are modern and up to date and permit widescale reconnaisance of groundwater sources.

A general observation of the regional features is always a first step in groundwater prospection. It is related to the vegetation, soils, geomorphology, available maps and aerial photographs. Particularly these latter are significant in the initial appraisal. As stated by Dijon [7], some characteristics traceable on the aerial photographs are indicators of the groundwater regime. For instance, the presence of Salsolacea can be related to the occurence of shallow salt waters, presence of furze and reeds is a sign that fresh water is present at shallow depth. Also increasing density of vegetation may indicate a buried river bed with some groundwater.

Also an appraisal of existing water localities, namely springs, ponds, oases, wells, etc. is made together with the appraisal of existing hydrological and water-quality observations, even if these are limited. Actually, in some cases when no advanced instrumentation is available, inventory of water points is a basic part of hydrological prospection.

The inventory of water points, containing information on the position, name, type, temperature, chemical quality, yield, results of pumping tests if any, and comments of the owner, has become a major part of many hydrological surveys working under the umbrella of water-resources departments, geological surveys, etc. Dijon [7] accounts for 7000 water points registered in Niger (from an estimated total of 20,000), 3000 in Mauretania, 2500 in Cameroon. Some 2000 points are registered in Zambia. The inventory becomes a basis for selection of systematically observed points and a part of country's water-resources development plan.

At present, the density of the observation points varies between one point for $4-400$ km^2. In populated areas the density is about $15-25$ km^2.

More frequently in Africa geophysical prospection has become an useful tool in hydrological prospecting. A method using resistivity measurement as a physical characteristic related to the lithology of the terrain and structure of the layers has been successfully applied to depths of 200 meters in regions where the results can be verified by drilling. The method has been used mainly in the sedimentary basins of Mauretania, Niger, Mali, Chad, Uganda, in the sandstones of Egypt, Sudan and Libya. Also in the weathered zones of the crystalline rocks in eastern and central coastal regions of Togo, Senegal and Ivory Coast.

The so-called electrical logging method, using a system of electrodes lowered into the boreholes to measure the upper and lower boundaries of the strata has been widely used in English-speaking Africa, particularly where a large number of boreholes are available within the studied region.

Seismic methods based on the observation of elastic waves produced by a shock are favoured in the shallow aquifers and limestones and crystalline areas. They have been used in Mauretania, Tanzania, Northern Cameroon, Ivory Coast, Chad, Togo, Upper Volta, Uganda, Dahomey, Madagascar and Mali.

The magnetic method, which measures the anomalies of the magnetic field, has been used for the detecting of tectonic features in the Rift region.

The application of isotope techniques has become a significant part of hydrogeological prospecting in Africa. In general, two paths are followed, one being based on injected radioactive tracers, another on the variation of environmental isotopes participating in the hydrologic cycle. The latter are mainly the stable isotopes deuterium and oxygen − 18 and tritium and carbon. The first three are parts of the water molecule. The fourth and other ones are present as dissolved isotopes and may be absorbed during any phase of the hydrological cycle.

Carbon − 14 has been used extensively for the dating of the groundwater in South Africa, mainly for the purpose of determining the age of water and thus the rate of recharge and the storage capacity of the aquifer in the arid regions of Southern Africa and Kalahari beds. The isotope is produced in the upper atmosphere and has a half-life of 5730 − 5568 years. It mixes with the atmospheric carbon-dioxide reservoir and as soon as it is not in contact with the reservoir, it becomes subject to radioactive decay.

The stable isotopes have been successfully used in several African projects. In the lake Chad area such an analysis indicated that the leakage from the lake to the groundwater is limited to the vicinity of the lake [11]. On the boundaries between Kenya and Tanzania the relationship between the lake Chala and spring regimes in the vicinity was studied and any possibility of a relationship was rejected because of the different stable isotopes composition [20].

Tritium with a half-life of 12.26 years is produced by cosmic radiation and by thermonuclear devices. Particularly in 1952 − 1963 the man-made pulses resulted in a strong labeling of waters by tritium, later on the methods based on tritium analysis have become more difficult in their application. Regarding its short half-life, tritium is recommended for the areas with rather quick exchange of water, which are more frequent outside the arid and semiarid areas. Tritium and stable isotopes were detected during the water supply project in Senegal. A great variety of isotopes was analysed in the ORSTOM project on the Lake Chad basin. Here significant results on the lake regime as related to the Bodele depression and Bahr-el-Ghazal basin were achieved.

Stable isotopes have been used in studying the Zinder area in Niger. Artifical radioactive tracers have been widely used in stream discharge measurement, bed material transport, sand movement and soil moisture, at many locations in the tropics and some of the instruments became standard tools of hydrological research in the tropics.

Drilling of boreholes is provided either as a final stage of the hydrogeological exploration or as a part of it. The boreholes for prospection are usually of a small diameter, however they may reach a great depth, say exceptionally 1000 meters. Some of the boreholes remain in function as controls of the variability of the water level during the operational stage. While the prospection boreholes may be of the size of 5−10 cm, in some cases when pumping tests are to be provided, the diameter is 10−25 cm, so that the pumps can be easily installed.

Boreholes destined for permanent pumping are even wider. The pumping tests are repeated at several levels and the duration of each test is 24−72 hours, for some purposes even longer. While the results of the pumping tests are reliable for sedimentary basins, in the areas with fissures, tectonic features and in karst, some difficulties may arise during the evaluation. Water quality measurements, concerned particularly with the measurement of dissolved solids and main ionic constituents, should accompany the pumping tests.

It should be pointed out that hydrogeological prospection is rather expensive and before starting any project a comparison of possible costs with available budget should be made. Reference is made to the budget analysis as performed by Burdon et al. [4].

On a continental scale, the groundwater potential of tropical Africa exceeds present and future needs. However, the unequal economic development within the regions and countries has led to the groundwater potential at some places being overexploited. This is particularly typical of areas with intensive agricultural and industrial development. In arid desert regions lack of groundwater, which is easily accessible or in a natural state, is another negative factor.

Many African capitals such as Abidjan, Lomé, Mogadishu, Nairobi, and Lusaka use groundwater as a major part of their water supply. The Economic Commission for Africa thought it economical to use groundwater for communities having less then 3000 people. The development of groundwater supplies is well justified in Africa, because of the structure of settlements. While the use of groundwater for the watering of cattle and other domestic animals and for industrial purposes is generally accepted, its use for irrigation should not exceed certain limits. In areas with a number of small-plot settlers, with say less than 2.5 acres, there is always a danger of depleting the water table by closely spaced boreholes.

In areas where surface waters are infected by tropical diseases and where the surface water supply requires construction of reservoirs, channels etc., the use of groundwater is more effective on any scale and some sources say it is up to ten times cheaper. The protection of boreholes against all types of pollution is quite simple. The maintenance of pumping machinery, particularly in remote areas or in the countries which do not produce machinery locally, may cause some difficulties. The same can be said for drilling machinery, which requires a reliable maintenance base having spare parts, not too far away.

Systematic groundwater survey is still inadequate except for some areas. International standards should be set for the establishment of a groundwater observation network with the aim obtaining a comprehensive picture of the groundwater resources of the African tropics. Prospecting in the crystalline rock areas of the tropics should be also extended.

Entirely different problems arise in connection with industrial development in the tropics from those related to the insufficient groundwater storage. For instance,

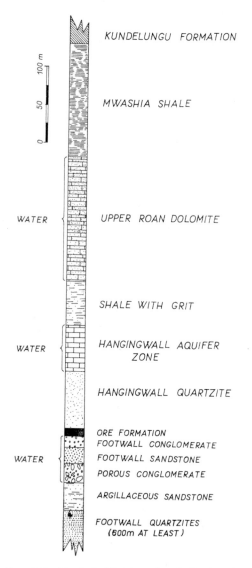

Fig. 5.7. Generalised section at the Konkola Division mine, indicating the most important water-bearing aquifers.

approximately 340,000 m³ of water has to be pumped daily to the surface from the Konkola mine of the Nchanga Consolidated Copper Mines on the Zambian-Zairean borders. A terrific excess of water makes this mine one of the wettest in the world. The water-bearing aquifer in the mine consists of conglomerates, sandstones and dolomites (Fig. 5.7). The profile also indicates how complete is the picture given by shallow drilling in certain areas and how extensive may be the groundwater resources still to be located elsewhere in Africa.

5.7 WATER IN SOIL

Soil water movement is another factor with an important role in the hydrological cycle of tropical basins, forming a bridge between the vegetation and the surface on the one side and groundwater aquifers on the other. According to several generally accepted classifications, soils of the same or similar types have been developed independently in various climatic zones and their classification does not coincide in general with the classification of the climate. Nevertheless, the climate is widely recognised as an important factor in soil formation. Thus the types of soils which once originated under tropical conditions are also characterized as tropical soils, though they are now located outside the limits of the tropics. Soils which have been formed under prevailing tropical conditions, in other words, under high aridity or humidity and high air temperatures are generally classified as tropical. Similarly found in the tropics are soils corresponding to nontropical climatic conditions. Long-term fluctuations of the climate, traced back to the Tertiary era, have played a significant role in the formation of the soil horizons under present tropical conditions. According to Kutílek [11] the role of climate still requires some re-evaluation, particularly when old sources of tropical classification are to be used.

Lang's factor defined as

$$L = \frac{S}{t}$$

where S means the mean annual precipitation in mm and t is the mean annual temperature in °C, has been frequently used as a source of information on the soil type where no better data are available. Here

$L > 160$ is typical for the process of podzolization,
$160 - 100$ for chernozem process,
$100 - 60$ for brown soil formation,
$60 - 40$ for laterites,
< 40 for desert soils.

In modern concepts, more attention is paid to the meso- and microclimates, particulary as related to the vegetation cover. The temperature and humidity regime is studied in addition to the parameters of Lang's factor. The wind regime is also another significant factor in arid parts of the tropics, while orography is a component playing an important role in the long-term process of soil formation. Finally, man's activity has begun to influence the upper parts of the tropical soils only recently. Fertilizing, overgrazing and cultivation of certain crops alter the soil rapidly.

African soils are formed from volcanic, granitic or sedimentary rocks with small zones of aeolian and alluvial sediments.

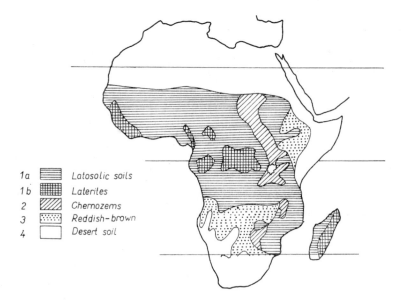

1a ▤ Latosolic soils
1b ▥ Laterites
2 ▨ Chernozems
3 ⊡ Reddish-brown
4 ☐ Desert soil

Fig. 5.8. Main soil types of tropical Africa.

Fig. 5.8 traces the four major soil types in tropical Africa. Laterites and latosolic soils are sometimes taken as a single group. They are chiefly located in the equatorial and savanna regions. They are heavily leached of salts and there is an accumulation of silica and oxides of iron, aluminium and manganese (Fig. 5.9a). Because of the rapidity of bacterial action, humus is entirely lacking. The soil is commonly a distinctive red in colour and is soft when first uncovered but becomes hard when exposed to the sun. It is poor, acid and lacking in mineral nutrients. Oxides are accumulated to such an extent that compact layers known as laterites are formed. Latosolic soils correspond with the equatorial climates and savannas.

Dark grey and black soils are sometimes classified as chernozems, although they are not as rich as those in Russia and America (Fig. 5.9b). They are usually developed on flat level plains and are associated with dark-coloured igneous rock. The black layer is less than 10 cm thick, rich in humus, and grades into a yellow-brown horizon. It is divided by a sharp line from a light-coloured horizon. The soil is rich in calcium appearing also as calcium carbonate. These soils require a more arid climate with hot summers and cold winters and also strong evaporation. Grasses flourish on them rather than forest. During heavy rains they are sticky and heavy and during dry seasons dry and cracked.

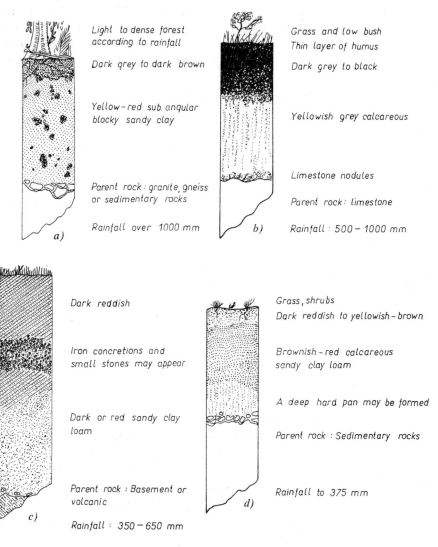

Fig. 5.9. Typical profiles of a) latolites, b) black soils, c) red loams, d) desert soils.

The third group is formed of red loams, chestnut and brown soils and prairie soils. Red loams are well developed on rolling land with a rainfall of over 1000 mm and a high temperature. The parent rock is granite, schist or sandstone. They are mainly found in savanna and they are not very heavily leached. Humus development is often good and they are fairly fertile. Chestnut and brown soils are developed in semi-arid areas. They are of a prismatic structure and contain less humus. When compared with chestnut soils, brown soils show a higher influence of aridity. Prairie soils are found less frequently in the tropics, in the transition areas between the red soils and black soils. They require over 700 mm of rainfall and are similar to black soils, lacking only calcium carbonate. A profile of red loam is shown in Fig. 5.9c.

Desert soils (Fig. 5. 9d) occur in a grey or red colour and contain little humus because of the limited vegetation. Horizons are very slightly differentiated. The profiles are shallow due to the absence of leaching by rainfall. Calcium carbonate occurs in the form of a lime crust, appearing as hard rock layer due to the slow evaporation of water near the surface. In depressions where there is no outlet for at least intermittent streams, the evaporation is intensive and the upper layer contains an excess of salt. Some of the soils are aeolian in origin and thus bear little relation to the rock over which they rest.

The water regime of the soils cannot be characterized by a more or less typical fluctuation for each soil type. Many additional factors, such as vegetation, slope, river network, cultivation etc. play a role and thus even neighbouring localities will behave differently as soil-water storage reservoirs. Several soil profiles belonging to group 3 on the map in Fig. 5.8 can serve as an example. In these profiles the soil moisture regime has been intensively studied for several years using gypsum block, neutron probes, and weighing methods [2], measurements being made down to the deepest groundwater level. Soil unit 63 consists of deep, well-drained, very strongly to strongly acid soils with moderately rapid permeability on $1-4\%$ slopes and with an estimated permeability of $6-12$ cm/hour. The soils have a surface layer of 2.5 cm of decomposed leaves and fine roots, which developes into a horizon consisting of dark, greyish-brown to brown sandy loam with a fine granular or subangular structure. This horizon may extend to a depth of 25 cm with a transitional underlay, extending to 50 cm and consisting of dark-brown to yellowish-brown or reddish-yellow sandy clay or clay with a fine structure and many fine pores. The colours are basically yellowish-brown or dark-brown with a few mottles of dark and red colour. This type of colour and texture extends below 3 meters. Fig. 5.10 gives a diagram of the annual soil moisture fluctuation within the profile down to a depth of 9 meters. The diagram shows the great variability of the soil moisture near the surface and the high stability of the soil moisture in the lower layers, which are saturated by a capillary rise.

Soil unit 83 consists of areas with a hard plinthite crust occurring either as outcrops or as an overlay of up to 100 cm of soil. The soils occur on gentle $2-4\%$ slopes. The soil has a 5 to 10 cm surface layer of very dark grey or dark greyish-brown sand

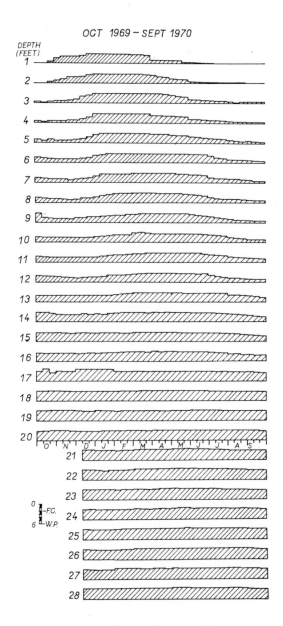

Fig. 5.10. Soil-moisture fluctuation in soil type 63. Parts of the deepest layers were not observed the whole year. FC — field capacity, WP — wilting point

142

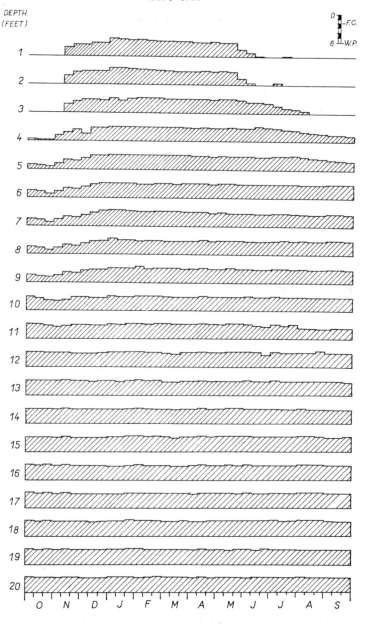

Fig. 5.11. Soil moisture movement in soil type 80.

or loamy sand. It is underlain by several inches of surrounded quartz and hard plinite gravel which rests directly on hard plinthite crust in the area of shallow soils. Above the stone line in deeper areas are loamy sands or sandy loams. Fig. 5.11 depicts a particular soil-moisture profile. The profile is depleted by woodland and, due to the slopes, the groundwater flows away instead of recharging the soil moisture of the deep layers from the bottom. The soil moisture movement is greatest in the upper layers and less pronounced in the lower parts.

Soil unit 60 is deep, poorly drained, slowly permeable, very strongly to strongly acid, with a water table during the dry season below 120 cm. The soil occurs on slopes of 2% to 3% and has a block surface of sand, loamy sand or sandy loam that

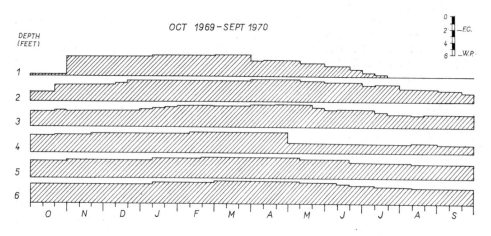

Fig. 5.12. Soil moisture movement in soil type 60.

varies in thickness from 12 to 25 cm, with very high organic matter content. With increase in depth, the colour becomes lighter. Root rust is frequently present. The texture of the soil may be sand, loamy sand, sandy loam or a coarse sandy-clay loam. The structure is weak, fine, subangular and blocky or single grain. At 55 to 120 cm there is a clear boundary with sandy clay or clay that has a weak to moderate, coarse or very coarse prismatic structure. The matrix colour varies from grey to white. Below 200 cm it grades into saprolite. Fig. 5.12 gives a diagram of the soil-moisture movement, where the soil is depleted by the transpiration from wet grasses. Between January and April the groundwater table practically reaches the soil surface, or there is a water layer on top of the soil.

5.8 LIST OF LITERATURE

[1] Anonymous, 1963. A review of the natural resources of the African continent. UNESCO Paris, 437 p.

[2] Aubrecht, G., 1959. Influence de divers types de vegetation sur les caractéres des soils en reg. equat. Unesco Symp. on Tropical soils, Abidjan.

[3] Balek, J., Perry, J., 1972. Luano catchments report. NCSR TR Report 28, Lusaka, Zambia, 70 p.

[4] Burdon, D., J., Caponera, D., A., Hrabovszky, J., P., 1971. The role of groundwater in social and economical development. The role of hydrology in the econ. dev. of Africa, Addis Ababa, 10 p.

[5] Bisset, C., B., 1941. Water boring in Uganda. Geological Survey of Uganda W. R. Papers, No. 1,. Entebbe, 32 p.

[6] Conrad, G., Foutes, J. Ch., 1970. Hydrologie isotopique du Sahara Nord Occidental. Symp. on Use of Isotopes in Hydrol., Vienna, 20 p.

[7] Dijon, R., E., 1971. Groundwater exploration. WMO Publ. No. 501, Geneva pp. 64—69.

[8] Dixey, F., 1962. Geology and geomorphology and groundwater hydrology. UNESCO Symp. on the Prob. of Arid Zones, Paris.

[9] Drouhin, G., 1953. The problem of water resources in Northeast Africa. UNESCO, Paris, 32 p.

[10] Dubief, J., 1953. Essai sur l'hydrologie superficielle du Sahara. UNESCO, Paris 32 p.

[11] Fontes, J., Ch., 1970. Deuterium et oygéne-18 dans les eaux du lac Tchad. Isotope Hydrology, pp. 387—402.

[12] Guidebook on Nuclear techniques in Hydrology, 1968. Technical Report No. 91, IAEA, Vienna, 213 p.

[13] Hellwig, D., H., R., 1973. Evaporation of water from sand. *Journal of Hydrology* **18**, pp. 93—108, 305—316, 317—327.

[14] Ivanovna, E., N., 1956. Essai de clasification géneral des soils. VIth Congress of ISSS E., Paris, pp. 393—401.

[15] Kellog, C., E., 1950. Tropical soils. IVth Cong. of ISSS E., Paris, pp. 247—253.

[16] Kutílek, M., 1963. Pedology of tropical and subtropical soils (In Czech), Prague Tech. Univ. Press, 107 p.

[17] Lambert, H., H., L., 1965. The groundwater resources of Zambia. Dept. of Water Affairs, Lusaka, 31 p.

[18] Martin, H., 1961. Hydrology and water balance of some regions covered by Kalahari sands in S. W. Africa. Symp. on Afr. Hydrol., Nairobi, pp. 450—455.

[19] Nace, R., C., 1969. Human use of groundwater. In "Water, Earth and Man", Methuen and Comp., London.

[20] Payne, B., F., 1971. Nuclear techniques in groundwater exploration. WMO Publ. No. 501, Geneva pp. 64—69.

[21] Payne, B., P., 1970. Water balance of Lake Chala, *Journal of Hydrology* **14**, pp. 47—58.

[22] Resources and Transport Div., UN, 1970. Groundwater in Africa. ECA Addis Ababa, 23 p.

[23] Schoeller, H., 1959. Arid zone hydrology. UNESCO Paris.

[24] Tisserout, J., 1956. Les resources en eau dans l'régions arides. Ann. P. Chaus., vol. 126, No. 3, pp. 965—997.

[25] Vogel, J., C., 1970. 14 C dating of groundwater. Symposium on the use of isotopes in hydrology, Vienna, 15 p.

6. AFRICAN LAKES

6.1 GENERAL

Most of the tropical African lakes are concentrated in the equatorial region where systems of tectonic rifts have developed. Favourable conditions for lake formation exist between the Zambezi river and the Red Sea by way of Tanzania, Kenya and the Ethiopian Highlands. In the present stage of development, many of the tectonic rifts are rather shallow depressions with muddy bottoms. Some are still in a complex stage of development.

The role which the African lakes play in the economy of the continent is important. They have always been one of the main sources food and the favourable conditions along the lake shores have been one of the positive factors in the development of African civilisation. Besides the natural lakes, new artifical lakes have been formed during the last few decades and still more are under construction. Small lakes and ponds seem to be more effective for the purpose of commercial fisheries than the big reservoirs.

6.2 NATURAL LAKES

The greatest African lake and the third largest in the world is the Victoria, known also as the Ukereve. The lake is 1135 meters above sea level. It is more than 400 km long, 240 km wide and covers an area of 66,400 km², according to some sources 69,480 km², depending on whether the swamps along the lakeshore are considered as part of the land or of the lake. The pan of lake Victoria is rather shallow with a mean depth of 40 meters and maximum of 80. The lakeshore is 7000 kilometers in length. Geologists assume that a much larger area was originally covered by the lake and Lake Kyoga was a part of it. The lake level was then 80 – 90 meters higher than it is today. Seasonal fluctuation of the lake is 65 cm on an average. On rare occasions, as in the sixties, a higher fluctuation is observed. The temperature range is rather small.

Lake Tanganyika is the second largest lake in Africa and the seventh in the world. The lake is 773 meters above sea level and covers an area of 33,000 – 34,000 km². Since the maximum depth of the lake is 1130 meters, the lake bottom is far below sea level. Tanganyika is, in fact, the second deepest lake in the world, next to Baykal. The average depth is 570 meters. The lake is 676 km long and 10 – 50 km wide. A subsurface ridge divides the lake into two pans. A number of short streams flow into the lake, the most important being the Ruzizi, flowing from Lake Kivu. The only outflow from the lake, the river Lukuga, is intermittent, in many years the lake evaporation exceeds the inflow. An outflow from the lake, the content of which is 19,000 km³, was made artifically in 1878 by digging through the banks at Albert-ville. Then the lake level decreased by 10 meters (up to 1892). Thus a volume of water equal to the capacity of two Kariba reservoirs flowed out from the lake. Since then, the long-term fluctuation is 2.5 meters and the annual fluctuation about 1 meter. The lake rise corresponds with the regime of the inflowing rivers, which means that it starts in December and continues till March (Fig. 6.1). With a view to all these facts, it can be concluded that Tanganyika is one of the natural lakes with the best natural balance, where inflow corresponds with evaporation. The evaporation, calculated by Thornwaite's formula, is 1696 mm, with maximum in September (167 mm) and minimum in February (124 mm). While the temperature in the first 480 meters has a slight variability, between 25−27 °C, from this depth to the bottom the temperature is fairly constant at 23.1 °C.

Nyasa lake is 472 meters above sea level. This lake is very long (579 km), but only 25−80 kilometers wide. The lake has an area of 28,000−31,000 km² and is the eleventh largest in the world. The mean depth is 273 meters and the maximum 706 meters. The volume of stored water is estimated at 7000−8000 km³. The lake is drained through the Shire river into the Zambezi.

Inflows are of no great size, the seasonal range of level fluctuation is 100 cm. As with lake Tanganyika, the shape of the lake encourages the formation of large waves.

At the margins of the rift formation, south of the Ethiopian Highland is located Lake Rudolph, 300 km long, 26−60 km wide, covering an area of 9000 km² and with a maximum depth of over 70 meters.

Lake Kivu is located in the zone of volcanic activity and covers an area of 2850 km. The lake is 96 km long and 48 km wide, with a depth of 485 meters and elevation of the level at 1460 meters. An interesting temperature stratification was observed at this lake. From a depth of 70 meters to the bottom, an increase of temperature has been measured, reaching 25.3 °C at a depth of 375 meters. This is explained by the influence of a warm, mineral-water inflow to the lower depths and by the contact of the lake bottom with the warm volcanic rock. The warmer mineral water remains stored at the bottom.

Lake Edward (length 80 km, width 48 km and an area of 2200 km²) forms, with the smaller George Lake, a system which drains through the Semliki river into Lake Albert (161 km long, 40 km wide, area 5300 km²). The lake system forms one of the important sources of the Nile. The shallow lake Kyoga, through which flows the Victoria Nile, and the Ethiopian lake Tana (1850 m. a. s. 1362.5 km²), the source of the Blue Nile, also belong to this system.

The only large tropical lake north of the equator is Chad. The lake is located

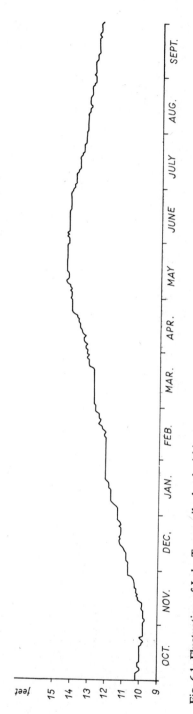

Fig. 6.1. Fluctuation of Lake Tanganyika level within a year,

in a shallow depression only 240 meters above sea level. Owing to the flat relief, the lake area varies greatly, depending on the amount of water brought in by the rivers Logone and Chari. Its size thus varies greatly between 10,000 and 25,000 km², while the mean depth is only 150 cm and the maximum only 12 m. The highest stage occurs in July when the lake increases in length to 250 km and in width to 150 km. The level then decreases up to December. The mean annual fluctuation of the level is below 80 cm and the maximum is 200 cm. Erosional processes result in the depression being flooded by eroded materials and the water becomes brackish. As the area is larger year by year, conditions are more favourable for increased evaporation.

Of local hydrological significance are lakes in the Sahara region, called shotts, which are very variable in size. Of larger size, but similar in behaviour, are the lakes in the pans at the margins of Kalahari, Etosha and Ngami. The Tanzanian lake, Rukwa, behaves in a similar way.

6.3 ARTIFICIAL LAKES

The natural landscape of Africa has been greatly changed during the past decades by man's activities. The construction of big dams is one of the activities which most changes the natural environment. The Volta and Kariba reservoirs, at least, should be mentioned in this connection.

Lake Volta, formed in 1962—66 on territory of Ghana, will probably long be one of the greatest man-made lakes. It covers an area of 8500 km² and it is 300 kilometers long. The lake is fed by the Volta river and is rather shallow and contains 148 km³ of water. Its maximum depth is 113 meters.

Lake Kariba was created after 1958 when Zambezi waters started to be impounded behind the Kariba dam. The lake fills a relatively steep escarpment. Beside the Zambezi, only small flash-flood streams contribute to the reservoir storage from Zambia; the rivers Gwaai, Sengwa and Sanyati contribute from the south. The lake is 320 km long, 20 km wide, with a mean depth of 29.5 and a maximum of 120 meters. It covers 5250 km² and contains 155 km³ of water. Owing to the growing economic interest in the lake, its limnological regime is perhaps better known than that of any other natural lake (Coche[1]).

In addition to these two lakes, Lake Koussou (Ivory Coast) covers 1700 km² and contains 29.5 km³ of water and Lake Kainji (Niger, Nigeria) covers 1250 km² and contains 150 km³ of water.

6.4 THE WATER BALANCE OF AFRICAN LAKES

The basic balance values, calculated for some of the lakes, vary considerably. According to the estimations of Hurst [2], Keller [3] and other authors, the following values were obtained (Tab. 6.1):

Tab. 6.1. Water balance of the African lakes

Lake	Area km^2	Inflow*) mm	Precipitation*) mm	Outflow*) mm	Evaporation*) mm
Victoria	66,400	241	1476	316	1401
Kyoga	1,800	3825	1270	3127	1968
Albert	5,300	4717	868	4151	1434
Edward	915	880	1360	800	1440
Nyasa	30,800	472	2272	666	2078
Tanganyika	32,890	1609	950	141	2418
Kariba	5,250	8440	686	7038	2088

*) related to the lake area.

While the values of the lake inflow and outflow can be measured with reasonable accuracy, the precipitation and evaporation are rather uncertain values. The rainfall values are normally available from the station along the lakeshore. The values of evaporation from various sources differ significantly. Mustafa's estimation [4] for instance, can be considered as reasonable information on the monthly fluctuation of evaporation from lake Albert (Tab. 6.2).

Tab. 6.2. Monthly values of the evaporation from Lake Albert

J	F	M	A	M	J	J	A	S	O	N	D	Year	
144	134	112	120	111	107	95	89	100	114	110	140	1376	mm

Even so, the annual values differ from the annual evaporation as given for Lake Albert in Tab. 6.1. Another example of difference in values with reference to Hurst's measurements can be given; this was partly made with the Piche evaporimeter and compared with the open-water evaporation measurements of the old data (Tab. 6.3).

Tab. 6.3. Hurst's measurements of annual evaporation

Place	Piche evaporimeter meters	Open water measurement meters
Oases	13.0	6.5
Khartoum	15.1	7.6
Albert L.	6.8	3.4
Edward	—	3.9
Victoria	—	3.9

Although the climatic conditions in the tropics appear to be more or less uniform from year to year and the lake evaporation independent of rainfall, there is a variability in annual evaporation. Thus, in the period 1957–64, the following values of annual evaporation were recorded at the station Samfya, on Bangweulu Lake (Tab. 6.4).

Tab. 6.4. Annual fluctuation of the evaporation from Bangweulu Lake

Year	Evaporation mm
1957/58	2492
1958/59	2445
1959/60	2069
1960/61	2116
1961/62	1913
1962/63	1860
1963/64	1726

Salinity measurements also show remarkable results, indicating high variability in salt content even with lakes belonging to the same system (Tab. 6.5).

Tab. 6.5. Salinity of some African lakes

Lake	Salinity ppm
Victoria	65
Kyoga	200
Albert	480
George	100
Edward	600

The various aspects of the African lakes were studied by the Worthingtons as early as in 1928. They found other remarkable differences in the fauna and flora, such as the presence of crocodiles in Lake Albert and the river Semliki, in contrast to Lakes Edward and George connected with the Albert through the Semliki, where they have never been found. More problems of this type should be solved by hydrologists, biologists and ecologists in close cooperation.

Perhaps the most reliable results on the water balance of Lakes Kyoga, Victoria and Albert can be found in the WMO report published after an extensive survey of the lakes and appertaining basins.

6.5 LONG-TERM FLUCTUATION OF LAKE LEVELS

The fluctuation of lake levels has been a matter of interest ever since the African lakes were discovered. The fluctuation of the water stored in the lakes reflects many other natural phenomena and, contrary to the fluctuation of rivers, can often be better traced to the past. Here the cooperation of geographers, hydrologists, geologists and archeologists must be emphasized. Plotted in Fig. 6.2 is the fluctuation of the Nyasa lake as compiled from the archives of the former Nyasaland Government. The levels are measured from the datum of the Shire Valley project, concerned

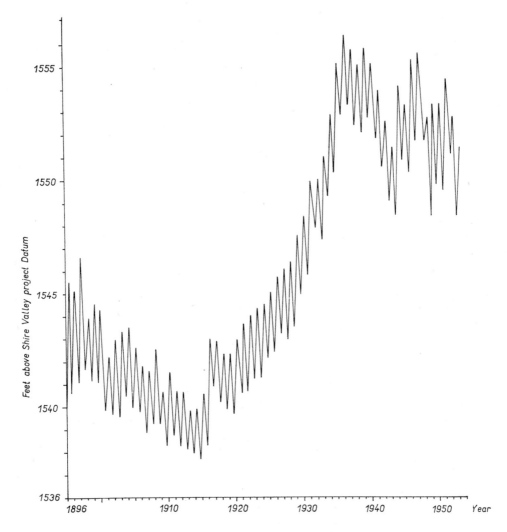

Fig. 6.2. Nyasa Lake, long-term fluctuation.

with the irrigation of the whole valley. When analysing such a record, one must take into account that an important role is played by climatological in addition to morphological factors. In the given case, it was the formation and destruction of the vegetational barriers at the outflow point. The compiled sequence shows the final part of the dry period in 1915 and continuous wet years since then, with maxima in 1937 and 1947/48.

The minimum and maximum levels of Lake Victoria (Fig. 6.3 and 6.4), as recorded

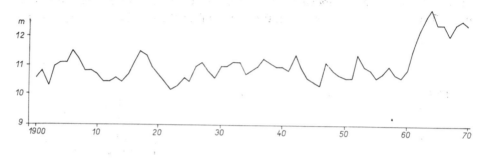

Fig. 6.3. Victoria Lake, annual maxima.

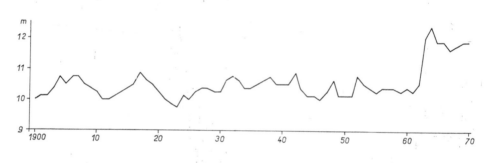

Fig. 6.4. Victoria Lake, annual minima.

by the Uganda Water Affairs Department, were analysed. The two diagrams plot the actual sequences and the sequences modelled from the most significant periodicities traced in them. The periodicities of 70.8, 35.5, 23.6, 11.8 years were found as most significant in the maxima sequence and 70.8, 35.5, 23.6, 17.7, 11.8 years in the minima sequence. A similarity with the periodicities traced in the rivers Nile and Niger (see Chap. 5) is remarkable. Also of interest in the sequences is a sudden increase commencingi n 1960 which probably occurred as a result of a change in the rainfall pattern.

Scientist concerned with tropical lakes are aware of the urgent need to obtain more information on African lakes. The outlines of a typical program in lake investigation were given by Coche [1]:

"Consequently the program for determining the average limnological conditions, over an entire reservoir should concentrate on the physical and chemical factors as, for example, temperature and dissolved oxygen. This information will lead to the identification of the various water masses and to the delimitation of sub-basin individualities where they exist. The limnological research program should perhaps be directed at first mostly towards understanding of the particular environmental factors related to the biology of fish stock."

6.6 LIST OF LITERATURE

[1] Coche, A., G., 1968. Description of physico-chemical aspects of Lake Kariba and impoundment. Fish. Res. Bul. Zambia 5: 200–267.

[2] Hurst, H., E., 1952. The Nile. Constable, London.

[3] Keller, R., 1962. Gewässer und Wasserhaushalt des Festlandes. Teubner, Leipzig, 520 p.

[4] Mustafa, I., 1970. Preliminary estimation of the water balance of Lake Albert. WMO Survey, Entebbe.

[5] Netopil, R., 1972. Hydrology of continents (In Czech), Academia Prague, 294 p.

[6] Sendder, T., 1965. Man made lakes and population resettlement in Africa. Symposium on man made lakes, London.

[7] Tabor, H., 1962. Solar energy. Problem of arid lands, Paris.

[8] Warren, W., M., Rubin, N., 1968. Dams in Africa. Cass and Comp. Ltd., London.

[9] Worthington, S. and A., B., 1933. Inland waters of Africa. MacMillan, London.

7. SWAMPS

7.1 ROLE OF SWAMPS IN AFRICA

African swamps, acting as natural water reservoirs, form an important part of the African economy. The total size of the tropical swamps has been estimated at 340 000 km², about one quarter of them being seasonal. According to Kimble [14], the total number of African swamps is estimated at between $10^4 - 10^5$. Some of the largest swamps are listed in Tab. 7.1. Although in general, the swamps are considered

Tab. 7.1. Main African swamps

Swamp	Country	Main stream	Area (km²)
1. Bahr el Jebel/Bahr el Ghazal	Sudan	White Nile	64,000
2. Middle Congo Swamps	Zaire	Congo	40,550
3. Lake Chad Swamps	Chad	Chari	32,260
4. Bahr Balamat	Chad	Chari	27,000
5. Okavango	Botswana	Okavango/Botletle	26,750
6. Upper Lualaba Swamps	Zaire	Lualaba	25,750
7. Lake Kyoga Swamp	Uganda	Victoria Nile	21,875
8. Lake Mweru Swamp	Zambia/Zaire	Luapula	17,000
9. Lake Mweru Wantipa Swamp	Zambia	Mofwe	16,750
10. Lake Bangweulu Swamp	Zambia	Chambeshi	15,875
11. Kenamuke/Kabonén	Sudan	—	13,955
12. Lotagipi	Sudan/Kenya	Tarach	12,937
13. Malagarasi	Tanzania	Malagarasi	7,357
14. Nyong	Cameroon	Nyong	6,688
15. Albert Nile Swamp	Sudan	Albert Nile	5,200
16. Kafue Flats	Zambia	Kafue	2,600
17. Lukanga	Zambia	Kafue	2,600

as a single ecological unit, from the hydrological viewpoint each swamp represents a special unit, depending on the bio-ecological, morphological and hydrological factors.

7.2 SWAMPS CLASSIFICATION

A unique feature of each swamp is its system of recharge and depletion. Until now no classification of tropical swamps has been established. On a world scale, in fact, swamp classification is rather different. The official glossary of the World Meteorological Organisation contains the definition from the English Dictionary which has been based more on historical terms and defines swamps, marshes and bogs jointly as "lowland flooded in the rainy season and usually watery at all times". Applied Hydrology Handbook differentiates between swamps and marshes on the one side and bogs on the other. Bogs are characterized as genetically related to lakes as a possible final stage of lake development, while swamps and marshes are defined as a vegetation-covered land area saturated with water.

According to Welch [19] swamps belong genetically to the "standing water series in which the water motion is not that of a continuous flow in a definite direction, although a certain amount of water movement may occur, such as internal current in the vicinity of the inlets and outlets".

Ivanov [13] analyzing the origin of swamps on the territory of the USSR, counted the following conditions as important for the formation of swamps:

a) flat relief and an impermeable soil or rock layer close to the surface,

b) clearance of the forest areas either by man or after a forest fire,

c) surface relief formed so that it can absorb the water excess from a much larger basin during the rainy season,

d) increased density of the vegetational cover on slowly flowing strams,

e) margins of river valleys where the groundwater aquifer is not capable of storing all the groundwater recharge.

Debenham [7] studied a variety of African swamps and defined them as products of those types of vegetational cover which tend to hold the backwater. Under such circumstances, the morphological conditions appear to be secondary, which, as will be shown later, is not fully correct. Kimble [14] differentiates between perennial and seasonal swamps. Under the latter heading can be listed a very special type of African swamp, the so called dambo. Kimble gives the following characteristics as being common to African swamps:

a) Runoff regulating systems acting in principle as reservoirs with an increased rate of evapotranspiration,

b) high ratio of surface area to water depth,

c) fluctuation in the size of the swamps from year to year and in some cases, from season to season,

d) three clearly marked zones: at the margin of the swamps is a zone under water for only a brief part of the year (usually at the end of the rainy season), the second zone is waterlogged for a much longer period and the third zone is under water throughout the year.

According to some authors, the seasonal and perennial swamps are of the same origin, both occurring whenever the morphological and climatological conditions are favourable to the waters of the rainy season congregating in a locality faster than they can disperse. They mainly differ only in size and in the role of the vegetation. There are examples of intermittent swamps in the Ethiopian Highlands lasting only several weeks and there are other examples of intermittent swamps in Zambia which may become fully saturated and perennial for several years.

7.3 SWAMPS AND DAMBOS

Intermittent swamps, called dambos, are thought by some authors to be genetically different from ordinary swamps. Ackerman [1] defined a dambo as ".. a streamless grassy depression periodically inundated and at the headwater of a drainage system in a region of dry forest or bush vegetation". His definition was based on his experience with dambos in the former Rhodesian territory. Hindson [10] defined a dambo as ".. seasonally waterlogged, grass covered treeless areas bordering a drainage line". According to Hindson, some dambos remain wet in the dry season largely due to the seepage which occurs along their margins and which results from slow subsurface drainage from the upland areas between the dambos. However, according to observations made in the Copperbelt [5], after a heavy rainy season continuing into the normally dry season, dambos may behave as if perennial, owing to their storage capacity.

A simple scheme gives an idea of the development of swamps and dambos in the tropics:

Lake (pond, stream)

   perennial swamp-intermittent swamp-land

Headwater area (lake, stream)

   perennial dambo-intermittent dambo-land.

Another difference can be traced to the source of the swamp recharge. Swamps are partially recharged by the precipitation falling on their surface, however, a stream normally serves as the main source. Dambos, on the other hand, are formed in the top parts of the channel network, where erosion cuts a valley where the rock is less weathered and covered by small deposits of soil. The rock is reached by erosion. The shallow location of the impermeable rock, contrary to the deeper location outside the dambo, may be caused by folding. This means that dambos, unlike ordinary swamps are recharged mostly by precipitation, since the subsurface inflow into them is relatively small and passes through the dambo into the drainage network, while the surface runoff from the dense bush cover is negligible. Thus a surface water layer, even if relatively thin, can be considered as a temporary stream draining the whole area. Another comparison is given in Fig. 7.1.

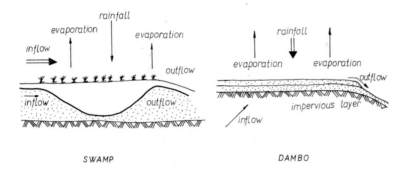

Fig. 7.1. Comparison of a swamp and a dambo.

Inside the dambo, the depth of the weathered rock is variable and the dambos usually have an assymetrical pattern. The diagrammatic section of the Luano dambos (after Gear [9]) in Fig. 7.2 gives an idea of the structure. In many cases, dambos are drained by relatively deep gorges and a relationship can be traced between them and stream development. Wherever there is a break in the dambo development, it is followed by a break in the tributary development and wherever there are small dambo outliers, they are drained by extensive tributaries. The drainage network is frequently hidden and can only be traced by using infrared photography. Obviously, the gorges are not a result of progressive headward erosion due to downstream changes in base level, but are rather the result of a change in the runoff characteristics of the dambos.

Dambo subsoils are usually poorly drained, slowly permeable, deep, strongly to very strongly acid, with the water table rather near to the surface. In many dambos, the lowest groundwater level is less than sixty centimetres below surface. They have usually been developed on a former Cretaceous erosional surface, and during the early Cenozoic era the surface was covered more extensively with a thick layer of deep, permeable and thick soils; the subsequent erosional cycles with plentiful rainfall caused the changes. Surface runoff therefore occurs when the groundwater storage space becomes overfull with excess groundwater.

Debenham assumed that dambos could be created artificially as they depend only on the establishment of a strip of dense vegetation holding the water longer in the dry season. Later, the strip would be extended more or less naturally. Thus, the possibility of unlimited development of dambos could be expected. It can be concluded from previous facts, however, that geological and geomorphological conditions play the most decisive role.

As stated by Hinton, a dambo can deteriorate from overgrazing and unprotected

158

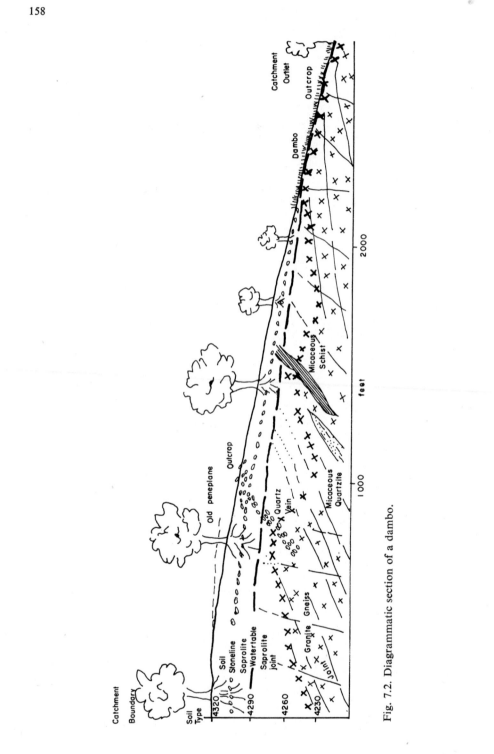

Fig. 7.2. Diagrammatic section of a dambo.

cultivation. Obviously, the formation of a dambo is a complex and slow process and overgrazing creates conditions for the formation of overland flow and sheet erosion, leading to a lowering of the aquifer. A similar effect can develop through unrestrained cultivation. Ditches draining a dambo may accelerate the groundwater outflow and reduce the size of the grassy area. The period during which the dambo releases water then becomes shorter.

With regard to overland flow, it should be pointed out that this occurs regularly every year in the dambo and forms a significant part of the total runoff. However, owing to the ability of the dambo vegetation to retain water, the overland flow is released slowly over two months instead of leaving the dambo area within a few hours. A separate hydrograph of a dambo outflow clearly indicates how the duration of the sruface runoff is prolonged until early June, whereas the last rainfall occurred in April (Fig. 7.3).

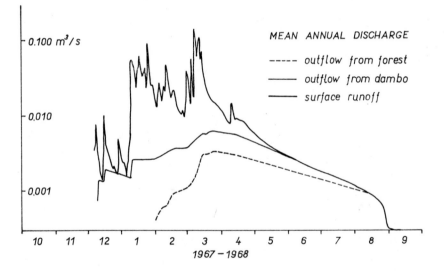

Fig. 7.3. Hydrograph of the dambo outflow indicating the duration of surface runoff.

The monthly distribution of surface runoff as a part of the total outflow is given in Tab. 7.2. Two dambos are compared in this table. In catchment B, the dambo forms 10.8% of the whole area, while in catchment J it is only 4.9%. While the total runoff remains approximately the same, the increased size of the dambo produces an increased volume of surface runoff, in this case, more than double. The seasonal distribution remains similar. The surface runoff occurs when the soil capacity of the dambo is exceeded or, in other words, surface runoff starts when an intermittent swamp is formed.

Tab. 7.2. Monthly distribution of the surface runoff from a dambo as a part of the total outflow

Month	Catchment B (1.13 km²)		Catchment J (1.28 km²)	
	Total	Surface	Total	Surface
	Runoff		Runoff	
	mm	mm	mm	mm
10	0.03	0.00	0.00	0.00
11	0.00	0.00	0.00	0.00
12	69.89	68.04	16.16	1.20
1	114.92	111.93	72.78	31.48
2	97.97	93.52	108.01	68.17
3	81.60	75.99	85.72	31.11
4	37.92	32.16	64.95	17.30
5	15.42	10.11	36.41	0.80
6	7.62	2.99	25.13	0.00
7	5.17	0.90	12.11	0.00
8	3.30	0.00	10.15	0.00
9	2.26	0.00	4.60	0.00
Annual	436.10	395.64	441.02	140.01

7.4 ROLE OF VEGETATION

Obviously, vegetation plays an important role in the hydrology of swamps and dambos. The genetic concept of swamps, as given by various authors, is related to the expected role of vegetation. Welch [19] related swamps to the so-called static environment or to the standing-water series. The vegetational concept, as suggested by Debenham, is based on the principle of running water, because aquatic vegetation requires water that is flowing. Therefore, according to Debenham, a swamp has to have an inclined surface and thus it can be said that the life of swamps is longer than that of the lakes, because the vegetation of swamps can adapt itself to many physical changes, other than continuous drought.

Apparently, both concepts are valid under certain circumstances; one can imagine a non-vegetation-covered intermittent swamp, even if the intermittency is related to a period of more than one year. Very little or no aquatic vegetation is found in the areas flooded once in several years, while in the seasonally inundated areas, the conditions for aquatic vegetation are very favourable.

The vegetation is obviously a significant element contributing to the hydrology of swamps. It produces permanent changes in the depth and direction of the channels inside the swamps and in the formation and floating of the islands, and these activities are sometime combined with the activity of animals. The hippopotamus in particular can influence swamp morphology, because when there are no hippos, the vegetation becomes denser and there are more opportunities for channel blockage.

Vegetation also characterizes the difference between swamps and dambos. The dominant plants of African swamps are *Cypherus Papyrus, Phragmites, Herminiera, Vossia suspidata, Pistia Stratiotes* and *Typha*. The conditions of frequently water-logged areas in the dambos are very favourable for a great variety of grass species. Verboom [18] listed fourteen species in the dambos of Zambia, Fanshawe [8] found more then sixty in four neighbouring dambos only. Dominant dambo grasses are *Andropogon laxatus, Aristida Atroviolacea, Brachiaria filifolia, Eragrostis capensis, Eriochrysis purpurate, Hyparrhenia bracteata, Hypogynium virgatum, Schizachyrium jeffroysi, Loudetia simplex*.

The main role of the vegetation is seen in its influence on the evapotranspirational process. The estimation of the evapotranspiration by various plants is still very approximate since it can be estimated only by indirect measurements, such as the discharge measurement of the inflow and outflow and rainfall measurements. Thus the role of the vegetational cover as a whole can be studied.

7.5 WATER BALANCE OF SWAMPS AND DAMBOS

There is a general presumption that the evapotranspiration from swamp vegetation is much higher than evaporation from a free water surface of the same size. Kostin [13] compared the evaporation from fully saturated soil with that from a free water surface and found the evaporation from fully saturated soil to be twice as high as that from free water. The experiment was carried out in moderate regions, but one can assume that in the tropics the difference can be even higher. Hurst [11] concluded that the evaporation from the Nile papyrus can exceed the evaporation from a free water surface. At that time, this was considered improbable. However, recent measurements from Uganda support Hurst's conclusions. The map of one section of the Bangweulu swamps (Fig. 7.4) shows a part of the swamps covered by papyrus. An evapotranspiration estimate was also made by comparing discharge and rainfall values at Bangweulu swamps [2]. Evaporation observed at a free water surface is 2340 mm and the evapotranspiration as calculated by water balancing is 890 mm. A mean annual precipitation of 1130 mm for the basin of the Chambeshi river flowing into the swamps has been determined and 1165 mm for the basin of the outflow, Luapula river. Correlating rainfall and runoff, an additional loss of 167 mm was estimated for the Luapula, which, when related to the area of swamps, represents 2000−2160 mm, according to the variability of the swamp area. Even the upper limit remains below free water surface evaporation. However, it must be taken into account that papyrus forms only a part of the swamps and that there are many islands inside them. Thus, about 60% of the total inflow is lost from the swamps by evapotranspiration.

A complicated hydrological regime exists in the Lukanga swamp, situated in the center of the Kafue basin, Zambia. The swamp was described by Macrae [16] and by Vajner [17], and its hydrological regime by Balek [3]. This swamp can be characterized as a sidestream reservoir. The river Kafue flows into it during the high flood season and passes it by in other months. Macrae believed that the channel connecting the river and the swamp can carry water in both directions and according to the FAO

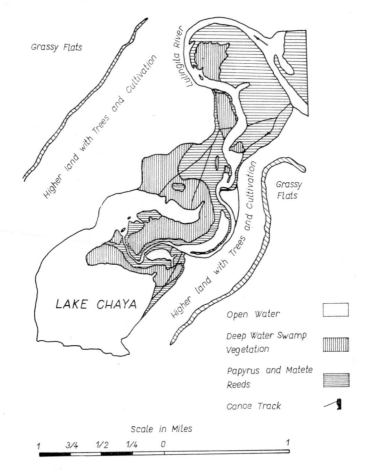

Fig. 7.4. Map of a part of the Bangweulu swamps. The entrance of Lulingila into lake Chaya.

survey, there is a possible spill of water during the flood season of more than 500 million cubic metres in an average year. The size of the swamp is approximately 2600 km^2 and the free water surface evaporation is 2070 mm. The annual precipitation above the swamp is 1250 mm and evapotranspiration 1020 mm. Within the basin drained directly into the swamp, the annual precipitation is much lower, only 970 mm. If the swamp were exerting no influence, the mean annual runoff would be at least 133 mm whereas actually, it is 123 mm. Thus an additional 10 mm approximately is lost in the Lukanga swamp.

Another water balance was calculated for a different type of swamp, the Kafue Flats. The swamps are located on both banks of the Kafue river, a tributary of the Zambezi. The swamps are saturated every year by the flooded river and dry out slowly during the dry season. Measurements taken on the free water surface indicate that 2070 mm of water are evaporated in a year. The mean annual precipitation for the whole basin, including swamps, is 1090 mm, the mean annual evapotranspiration from the Kafue basin is 986 mm. The evapotranspiration from the swamps is 1005 mm; exluding the swamps, it would be only 814 mm. Thus, an overall loss of 4% of the total inflow is caused by the swamps, however with a view to the abundant game living in and near the swamps, the term "loss" does not seem to be very accurate.

Tab. 7.3 compares the results obtained from the water balance with the hydro-

Tab. 7.3. Results of water balance calculations for some swamps and dambos

Parameter	Unit	Bangweulu swamps	Kafue Flats	Lukanga	Dambo
Drainage area	km^2	102,000	58,290	19,490	1.43
Area of swamp	km^2	15,875	2,600	2,600	0.15
Rainfall on the area in a year	mm	1,190	1,090	1,250	1,330
Rainfall on the swamp in a year	mm	1,210	1,110	970	1,330
Evaporation from free water surface, yearly	mm	2,340	2,070	2,070	1,710
Evapotranspiration outside the swamps, yearly	mm	890	785	908	1,320
Additionally evaporated from the swamps, yearly	mm	1120—1260	196	252	—
Total evapotranspiration in the swamps, yearly	mm	2000—2180	1,000	1,120	1,075
Lost in % of inflow	%	60	4	7.8	—

logical effects of a dambo. Obviously, the effect of the swamp is very variable, depending on its type, and particularly its structure and vegetation. The Bangweulu swamp really functions as a reservoir, transforming the hydrological regime of the Chambeshi, the ultimate upstream of the Congo river, into the different regime of the river Luapula. The Kafue Flats have a channel system which is much less developed than that of the Bangweulu. Lukanga is an example of a side swamp with

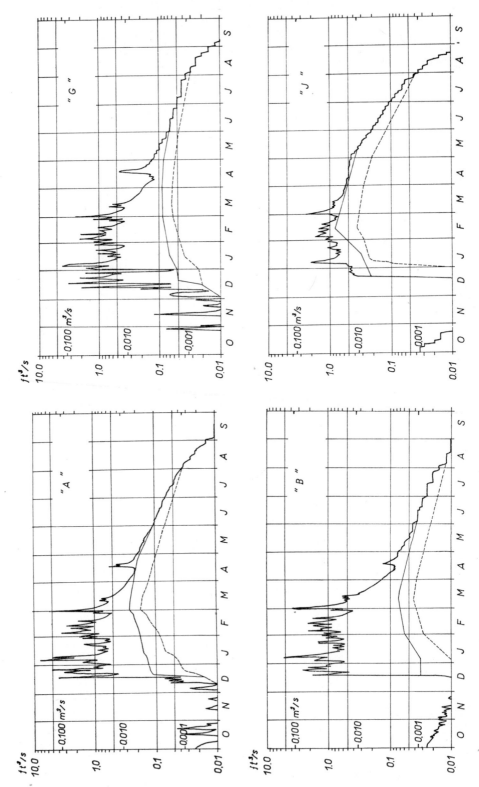

Fig. 7.5. Separate hydrographs of four different dambos.

limited connection with the main stream, however, it can accept any amount of flood water flowing from the headwaters of the river. The dambo is surrounded by dense forest formed mainly by the Brachystegia species, which can easily evaporate any amount of water infiltrating into the soil. The storage capacity of dambos is very limited and so the volume of outflow from them is more or less constant year by year and only varies slightly at the beginning of the rainy season, before the dambo storage zone has been completely saturated, depending on the distribution and intensity of the initial rainfall.

Fig. 7.5 gives separate hydrographs of four different dambos. The lowest part indicates in each case the basis flow from the so-called transitive region at the margin of each dambo. The vegetation there is a mixture of grass and poor woodland, varying in size according to the morphological conditions, the geology of the area and the vegetation cover. In some cases, all the area surrounding the swamp may become transitive. The middle part of the hydrographs represents the contribution from the dambos, while the top part is the surface runoff or the outflow from the dambo surface. The surface runoff does not occur as a direct product of rainfall. It is a product of the overstored groundwater zone. Instead of a steep fall in the surface runoff hydrograph, a slow decrease is typical for these parts of the hydrographs. There is also a remarkable difference in the shape of all four hydrographs, although all are located in the same neighbourhood. Their basic characteristics are as follows (Tab. 7.4):

Tab. 7.4. Basic characteristics of four dambos in the Copperbelt

Dambo	Dr. area catchment km^2	Dr. area dambo only km^2	% of total area	Shape
A	1.43	0.147	10.3	square, equal sides
B	1.13	0.122	10.8	rectangular, elongated
G	0.94	0.102	10.8	triangular, equal sides
J	1.28	0.058	4.5	triangular elongated

Fig. 7.6 shows the influence of the size of the dambo area, in relation to the total size of the whole catchment, on the surface runoff. The smaller the dambo, the lower the surface runoff.

A comparison of the seasonal distribution of evapotranspiration from a dambo and a swamp is also interesting. The comparison was made for the Bangweulu swamps and a dambo in the Copperbelt area. Both localities are within a distance of 160 km of each other, but their evapotranspirational effects are entirely different (Tab. 7.5):

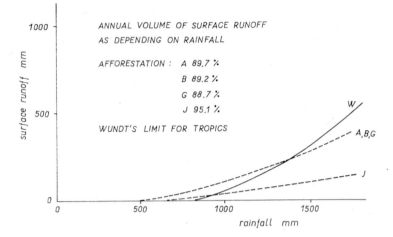

Fig. 7.6. Influence of dambo size on surface runoff.

Tab. 7.5. Evapotranspiration from swamp and dambo

Month	Dambo		Swamp	
	mm	% of annual total	mm	% of annual total
10	28	5.7	114	5.3
11	90	18.2	261	12.1
12	94	19.1	249	11.5
1	97	19.7	250	11.6
2	69	14.1	261	12.1
3	53	10.7	276	12.9
4	26	5.4	208	9.6
5	10	2.0	122	5.8
6	9	1.8	97	4.4
7	6	1.2	101	4.6
8	5	1.0	107	5.0
9	5	1.0	109	5.1
Year	494	100.0	2156	100.0

The figures for the months of January to September remain more or less the same from year to year, however, the first three months of the hydrological year may have a surprisingly different evapotranspiration each year, depending on the intensity and time distribution of the initial storms. The main difference, as observed in the table, is at the end of the dry season when the water storage in the dambos is at its lowest value, while in the swamps there is still enough water for evapotranspiration.

Although the role of the swamps in the hydrological regimes is not fully known, some observations from the Central African Plateau indicate how the presence of swamps reduces the surface runoff. The values as given in Tab. 7.6 are very much smoothed and deviations can be expected from case to case, even so, they can serve as a preliminary estimation.

Tab. 7.6. Annual runoff as depending on the size of swamps and annual rainfall

Areal extent of swamps as a percentage of the drainage area, %	Annual runoff (mm) at annual rainfall of		
	1250	1000	750 mm
1	358	182	56
2	315	155	41
3	275	127	31
4	243	107	23
5	220	91	20
6	203	76	15
7	188	71	10
8	177	68	8
9	172	66	5
10	167	63	3

As has been proved experimentally [4], outflow from the swamp and the accumulated water storage are not only in relation to the annual precipitation in the same year, but previous years frequently play an important role, and sometimes precipitation over as many as four years before was found to be significant. Equally interesting results were found for water-balance values on a monthly basis. For the mean water level of April, the monthly rainfall total in September was significant in the southern hemisphere and on the contrary, mean water levels in the rainy season had no relationship to the mean water levels in previous months.

One of the first attempts to simulate the water regime of the large African swamps was made by Hutchinson and Midgley [12]. A complex of the river and swamp system in the lower reaches of Okawango, Botswana, was simulated using in principle the Muskingam method. Severe earth tremors in 1952 produced changes in the gradient of the swamps which made the simulation rather difficult. The authors felt that ecological studies should be undertaken in swamps before further progess in simulation could be achieved.

7.6 LIST OF LITERATURE

[1] Ackermann, E., 1936. Dambos in Northern Rhodesia. Wiss. Veroff. Leipzig, 4, 1 p. 149—157.

[2] Balek, J., 1970. Water balance of Lake Luapula and Lake Tanganyika basin. NCSR Rep. TR 10, Lusaka, 15 p.

[3] Balek, J., 1971. Water balance of the Zambezi basin. NCSR Rep. TR 15, Lusaka, 20 p.

[4] Balek, J., 1972. Water level forecasting at the Bangweulu Swamps. NCSR Working Paper, Lusaka, 6 p.

[5] Balek, J., Perry, J., 1972. Luano Catchment Project, First Phase. Nat. Council for Scient. Research. Tech. Rep. 15, Lusaka 20 pp.

[6] Balek, J., Perry, J., 1973. Hydrology of African headwater swamp. *Journal of Hydrology* **19**, 1973, pp. 227—249.

[7] Debenham, F., 1952. Study of an African swamp. Colonial Office London, 52 p.

[8] Fanshawe, D., 1971. The vegetation of Luano catchments. Luano Catchments Research Project, NCSR Rep. WR 10, Lusaka pp. 36—47.

[9] Gear, D., 1968. Technical Annex to the final report on the representative Luano catchments. NCSR unpublished rep., Kitwe, 177 p.

[10] Hindson, J., R., E., 1955. Protection of dambos by means of contour seepage furrows. Minist. Agric. Int. Rep., Lusaka, 53 p.

[11] Hurst, H., E., 1954. Le Nil. Payot, Paris.

[12] Hutchinson, I., P., G., Midgley, D., C., 1973. A mathematical model to aid management of outflow from the Okawango swamp. *Journal of Hydrology* **19**, pp. 93—112.

[13] Ivanov, K., E., 1953. Gidrologiya bolot (Hydrology of swamps), (In Russian), Gidro-meteoizdat, Leningrad, 294 p.

[14] Kimble, G., T., 1960. Tropical Africa. The Twentieth Century Fund., New York.

[15] Kostin, J., I., 1954. Meteorologiya i klimatologiya (in Russian), Gidrometeoizdat, Leningrad.

[16] Macrae, F., G., 1934. The Lukanga Swamp. Geograph. Journal, London, Vol. 83, pp. 213—218.

[17] Vajner, V., 1969. The Geology of the Mumbwa Areas. Geol. survey, Lusaka.

[18] Verboom, W., C., 1965. Aerial photographs, relics of soil, vegetation and tse-tse flies. I. T. C., Delft, 21 p.

[19] Welch, I., S., 1952. Limnology. McGraw Hill, New York.

8. HYDROLOGICAL EXTREMES

8.1 GENERAL

A general view of the occurrence of extremes under tropical conditions is somehow different from the view taken under moderate conditions. Floods, usually considered as disasters elsewhere, are taken as a sort of blessing under arid conditions. They are expected and awaited before they arrive. In some regions, the fertility of the land depends on the maximum stage of the flood. The water tanks of arid regions are recharged only from surface runoff and torrential floods. New dams are also recharged by the flood regime. Dry periods in the regions of wet and dry climate are also anticipated and the agricultural routines, crops, herds and game are well adapted to it. Nevertheless, there are occasions when the dry period is too prolonged, or there is not enough rain between two or more dry seasons. An exceptional flood, covering large areas with a sheet of water, comes once in several years and under detailed analysis, one can see that the general probability law of the occurrence of hydrological extremes is more or less the same as that valid elsewhere.

Only a few rivers in Africa have been observed for at least fifty years and the numerous gaps in the records make it difficult to analyse the sequences of extremes properly for the needs of engineering designers. In some cases a hydrologist in the field depends on oral records which are usually not very reliable, but which at least give some basis for a sound estimation. For instance, when tracing an exceptional flood in the Luangwa river basin, Zambia, information from the local senior chief that "nothing like this flood has been witnessed in the valley the whole time the tribe has been living there" was quite reliable for estimating that the flood was at least of a 1% probability of occurrence, since the tribe had lived in the valley for about 100 years.

8.2 FLOODS

Flood studies in African countries were initiated and carried out by French teams, particularly by the ORSTOM. Several experimental basins in west Africa had been established for that purpose already before the International Hydrological Decade started and many results related to each particular basin were published.

Attempts were made right from the start to establish the relationship between effective rainfall and the formation of hydrographs. Rodier [8] studied in the Sahel region the basic characteristics of the effective rainfalls producing flood hydrographs.

A difficulty in carrying out such a study lies in the non-availability of autographic records; normally, only daily totals are available. Therefore, in accordance with the estimation of permeability, the limits of infiltration conditions were established which when combined with the daily totals, gave basic information on the rainfall-runoff relationships. For impermeable soils an infiltration value of 10 mm/hour and for permeable soils a value of 40 mm/hour were estimated; in the regions with an annual rainfall of 300–1000 mm, it was calculated that effective rainfall, up to 85% of the daily total, can fall in 90 minutes on soils accepting 20 mm/hour and 75% of the daily total can fall in 55 minutes on soils accepting 40 mm/hour. Providing the rainfall record is long enough, such a study can serve as a basis for the calculation of effective rainfall with a certain probability of occurrence.

Tab. 8.1. List of floods on Pra river, Ghana, 1944—1960, one flood per year included, empirical probability calculated

No.	Date		Q max m^3/s	$p = \dfrac{m}{n+1} \cdot 100^*)$
1	7	60	1280	5.57
2	7	57	1020	11.14
3	7	53	990	16.70
4	9	47	810	22.30
5	11	55	780	27.85
6	6	58	780	33.40
7	6	56	744	39.00
8	10	51	720	44.50
9	7	44	660	50.10
10	7	49	660	55.70
11	9	52	570	61.20
12	10	59	550	66.80
13	10	45	504	72.40
14	6	48	504	78.00
15	7	54	504	83.50
16	10	46	420	89.00
17	10	50	262	94.60

*) m = number of event

n = total number of events

In the initial studies carried out in Southwest Africa, the flood regime was studied with regard to the number of years in which the river flows reach the ocean. The River Swakop, as described by Wipplinger [11], flooded once in six years, as did the River Gamams. According to the author, a certain characteristic of the flood regim is the marked degree to which flood waters are dissipated on their path down into the sandy channels, because the flow of the river into the sea occurs very seldom owing to the dissipation. This is, of course, a very special approach to the flood frequency analysis, influenced by extremely arid conditions and by the ephemeral regimes of the wádí type of the rivers. However, in arid regions, a similar analysis

Tab. 8.2. List of floods on Pra river, Ghana, 1944—1960, all floods included

No.	Date		Q max m^3/s	$p = \dfrac{m}{n+1} \cdot 100$
1	7	60	1280	2.86
2	7	57	1020	5.70
3	7	33	990	8.60
4	10	60	840	11.40
5	9	47	810	14.30
6	11	55	780	17.10
7	6	58	780	20.00
8	6	56	744	22.80
9	10	51	720	25.70
10	7	55	672	28.60
11	7	44	660	31.50
12	7	49	660	34.30
13	10	49	635	37.20
14	10	53	605	40.00
15	9	52	570	42.80
16	6	52	550	45.70
17	10	57	550	48.50
18	10	59	550	51.40
19	5	59	505	54.40
20	10	45	504	57.20
21	6	48	504	60.00
22	7	54	504	62.80
23	11	54	460	65.20
24	11	55	436	68.60
25	5	59	436	71.40
26	10	46	420	74.20
27	10	44	380	77.20
28	10	56	366	80.00
29	11	48	282	82.80
30	10	50	262	85.60
31	7	45	250	88.70
32	6	50	250	91.40
33	7	46	232	94.20
34	10	58	196	97.20

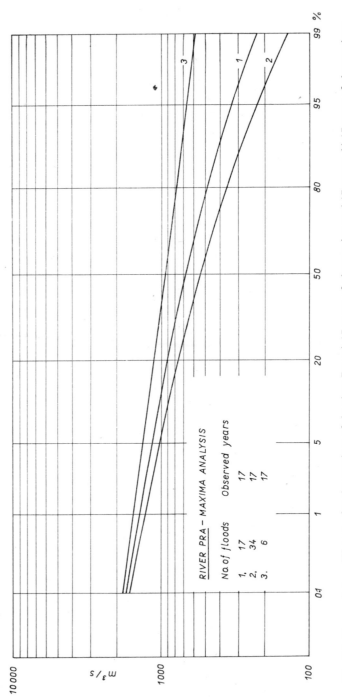
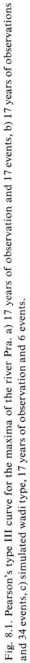

Fig. 8.1. Pearson's type III curve for the maxima of the river Pra. a) 17 years of observation and 17 events, b) 17 years of observations and 34 events, c) simulated wadi type, 17 years of observation and 6 events.

can be undertaken not only for the mouths of the rivers, but for any other cross section. Beside the occurrence of the event, its length can also be used as an additional characteristic, when the actual amount of the discharge in sandy river beds is unknown. Thus the season of 1934/35 was considered as particularly important for the Swakop river, because the river then flowed continuously for four months.

In many formulas calculating a 1% flood, a simple assumption is made that the river flood occurs once a year; when analysing the flood regimes of some West African rivers [3] a simple approximate formula was developed reflecting the possibility of the occurrence of more floods in one year, or in other words, an unequal number of years and events. This is particularly useful for rivers with a double flood regime in a year and also for the ephemeral streams. Tab. 8.1 gives a sequence of seventeen culminations of the river Pra, Ghana. Here only one maximum per year was taken into account and the empirical probability curve calculated, In Tab. 8.2 the sequence of all floods observed within the same period was analysed and both empirical curves plotted together in Fig. 8.1. The probability scale was reevaluated by using the formula

$$p = \frac{100}{N \dfrac{n}{M}} \%$$

where p is the probability of the occurrence of the flood being repeated once in N years, n is the number of events and M is the number of years of observation.

The results are plotted in Fig. 8.1 and the numerical results given in Tab. 8.3. Here we can see how the results differ for the highest and lowest probabilities.

Tab. 8.3. Floods as calculated for various probabilities and number of events, period 1944—1960

1 flood in a year		2 floods in a year		6 floods in 17 years	
N years	Q m^3/s	N years	Q m^3/s	N years	Q m^3/s
100	1710	100	1670	100	1700
50	1520	50	1510	50	1480
20	1300	20	1290	20	1180
10	1110	10	1120	10	1050
5	900	5	960	5	900
2	610	2	715	2	—

As an additional example, a hypothetical sequence was analysed simulating an ephemeral flood stream from the basic sequence in Tab. 8.1. Only the first six floods were considered as occurring during the seventeen-year period. The empirical and hypothetical curves are plotted in Fig. 8.1 and Tab. 8.3.

Because not even short sequence are yet available for many African streams, regional formulas were developed for some regions. Pitman and Midgley [7] applied Hazen's method and the Myer-Jarvis formula to the conditions found in South Africa and divided the country into seven regions:

1. High rainfall areas, mountains, a high proportion of exposed rocks, sclerophyllous vegetation.

2. Interior plateau, relatively high rainfall, with early summer thunderstorm belt, grassveld, fairly dense drainage pattern, parts suffering from advanced soil erosion.

3. Year round rainfall, dense drainage pattern, high proportion of exposed rocks or advanced soil erosion.

4. Eastern coastal strip, high rainfall, dense drainage pattern, generally well vegetated.

5. Subhumid to semiarid interior plateau, rains generally late summer.

6. Well-forested bushveld areas, low rainfall areas well forested.

7. Arid interior and west coast, sparse vegetation, low drainage density.

Although only zone 6 is located in the tropics, the results as supplied in Tab. 8.4 can be extended north of the Tropic of Capricorn, provided the similarity of the regions and regimes has been proved and the mean annual flood peak regime is known.

Tab. 8.4. Flood peak/mean annual peak ratio for various recurrence intervals and South African regions

Region	Flood peak/mean annual flood peak for recurrence intervals (year) of				
	5	10	20	50	100
1	1.54	2.08	2.67	3.54	4.28
2	1.70	2.43	3.25	4.48	5.54
3	2.00	3.14	4.51	6.68	8.60
4	1.97	3.02	4.24	6.13	7.78
5/7	2.04	3.22	4.58	6.82	8.78
6	2.46	4.40	6.82	10.75	14.26

For direct application, reference is made to the coaxial diagram as plotted in Pitman and Midgley's paper.

The mean annual flood, as related to the catchment area, was analysed by Kovács [5] for some East African catchments. The author compared the water yield of mean annual floods with the catchment area and differentiated between the catchments of limited area, that is, catchments influenced by heavy rainstorms of high intensity on small areas, and catchments of large size under the influence of less intensive rainfalls but on a more extensive area. As indicated by Fig. 8.2 the difference in length of the horizontal part of the curve is the main characteristic of each particular regime.

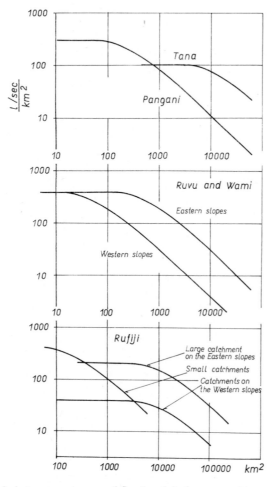

Fig. 8.2. Relationship between mean annual flood and drainage area for some East African rivers, after Kovács.

A flood analysis of Kenyan, Malawian and Nigerian rivers was provided by Starmans [9] who developed the graphical relationship between the drainage area and the so-called peak discharge, a term related to a 1% flood (Fig. 8.3). For a short period of observation (1950–1971) a study of high and low water years was undertaken by Starmans and Shalash [10].

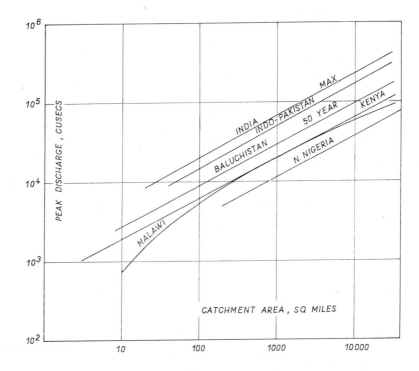

Fig. 8.3. Relationship between the drainage area and peak discharge for some Kenyan, Malawian and Nigerian streams, after Starmans.

A comparison between the 1% flood as calculated for some African basins and South African and European 1% floods was calculated by Balek [2]. An analysis of several flood sequences was provided using Pearson's type III curve [6]. The results are found in Fig. 8.4 and the validity of the curves is described in Tab. 8.5.

Tab. 8.5. African and European regions having different flood regimes

1. Insufficiently afforested or denudated areas of the Mediterranean.
2. Mountainous areas and deep valleys of the Central African Plateau.
3. Mountainous regions of South Africa.
4. Mountainous regions of Central Europe.
5. Upper reaches of the Zambezi, Congo, headwaters. Streams with pronounced rolling character of the basin, not influenced by swamps.
6. African West Coast.
7. Tropical rivers with flat or less pronounced rolling character of the basin, not influenced by swamps.
8. Flat basins of Central Europe.
9. Flat tropical basins influenced by swamp regimes.

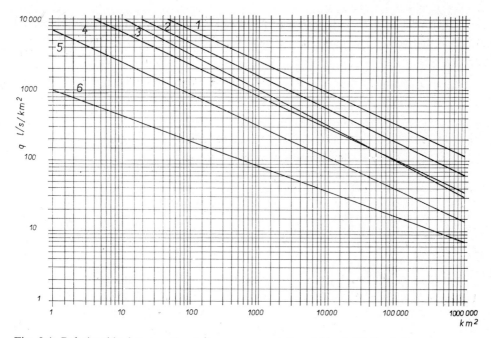

Fig. 8.4. Relationship between the drainage area and the yield of 1% flood for African and European streams.

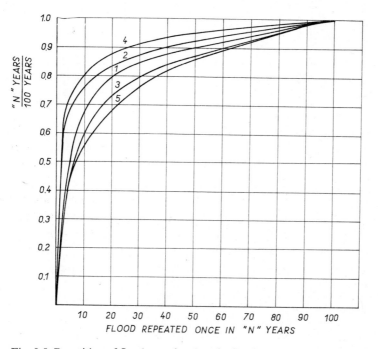

Fig. 8.5. Repetition of floods as related to the flood repeated once in 100 years.

Fig. 8.5 serves for a calculation of flood repetition as related to a flood repeated once in 100 years. The curves are typical for the regions as described in Tab. 8.6.

Tab. 8.6. Regions for which are characteristic the curves in Fig. 8.5

1. Upper Congo reaches.
2. Rolling basins of African Plateau.
3. Mountainous basins and escarpments.
4. Undulating basins of Central African Plateau.
5. Flat tropical basins.

8.3 DROUGHTS

Droughts in the tropics produce perhaps more disastrous situations than floods. Unfortunately, from the hydrological point of view, the effects and statistical analysis results are far less known than for floods and prevention of their effects is much more difficult than the effects of floods. The most significant difference probably lies in the duration of the extremes. While the flood peak is a matter of several hours, perhaps days, the maximum effects of a drought may continue for weeks, months and sometime even years. Under a moderate climate and with perennial rivers, droughts can be defined by the lowest stage of a river at a certain cross section; under an arid climate, however, there are many rivers which regularly dry out. The duration of the intermittent period then becomes the most significant characteristic. It is natural that many tropical rivers dry out regularly for several months and that only a prolonged dry period is considered as catastrophic, while the catchments are ecologically well adapted for the normal dry period.

The regime of the perennial rivers can be characterized and calculated as for maxima. Fig. 8.6 shows the Pearson's type III curve, calculated for the sequence of minimum discharges of the river Ubangi, indicating that the lowest recorded discharge of 344 m³/s was less than the 1% minimum. The minima analysis of intermittent streams is simplified, since the minimum discharge is always zero; when studying an intermittent stream, instead of the discharge values, we analyse a sequence of the lengths of the dry periods, just as for the flood sequence. The sequence of the appertaining mean annual rainfall totals can be also used for such an analysis.

One of the worst droughts in Africa was experienced during the period 1968–1974 in the Sahel region. Chad, Mali, Mauretania, Niger, Senegal, Upper Volta and Gambia suffered most. The drought extended to southern Sudan and Ethiopia (Fig. 8.7). During the dry period, the rivers Senegal and Niger fell to the lowest stages measured and Lake Chad evaporated to only one third of its normal size.

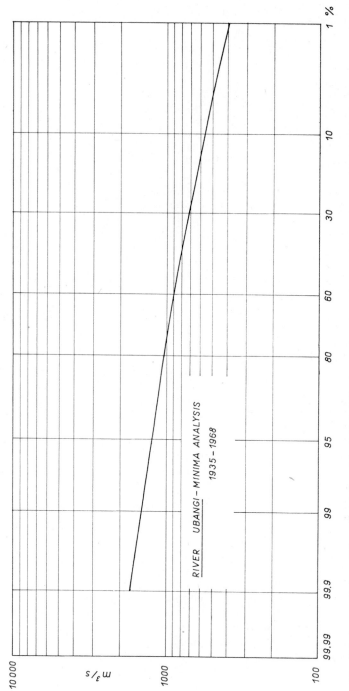

Fig. 8.6. Pearson's type III curve and empirical probability curve for the minima of the river Ubangi, Central African Republic.

180

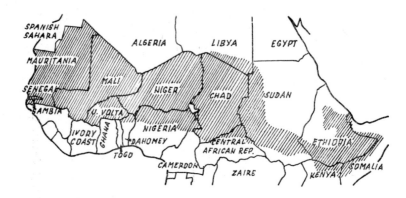

Fig. 8.7. Region suffering from a catastrophic drought in 1968—1974.

The arid zone at some locations extended 100 km/year southward. The soil infiltration also decreased due to the poor quality of the soil and so when the heavy rains came in 1974, most of the water flowed away as surface runoff, causing additional damage to the population and crops.

8.4 LIST OF LITERATURE

[1] Balek, J., 1972. An application of inadequate data for the Afr. trop. regions in engineering design. 2nd Int. Symp. on Hydrology, Fort Collins, Col., 8 p.
[2] Balek, J., 1971. On extreme floods in Zambia. NCSR WR. Report 11, Lusaka, 16 p.
[3] Balek, J., Holeček, J., 1964. Contr. to the calculation of the flood probability (In Czech). *Vodní hospodářství* 4/64 pp. 83—85.
[4] Klein, J. C., 1964. Étude hydrologique de bassins expérimentaux dan L'Est Volta. Min. de l'Écon. Nat., Haute Volta, 77 p.
[5] Kovács, G., 1971. Relationship between characteristic flood discharges and the catchment area. Symp. on the Role of Hydrol. in Ec. Dev. of Africa, Vol. II, WMO No. 301, Genéve, pp. 18—23.
[6] Němec, J., 1972. Hydrology for Engineers. McGraw Hill.
[7] Pitman, W., V., Midgley, D., C., 1967. Flood Studies in South Africa. Die Siviele Ingenieur in Sid-Afrika, 1967, pp. 193—198.
[8] Rodier, J., 1961. Essais de determination des caracteristiques des averses décennales. Hydrol. Symp. Nairobi 1961, pp. 121—126.
[9] Starmans, G., A., N., 1972. Personal communication, Lusaka, Water Affairs Dept.
[10] Starmans, G., A., N., Shalash, S., 1972. High and Low Water Years in Zambia. Min. of Rural Dev., Lusaka, 100 p.
[11] Wipplinger, O., 1961. Deterioration of catchment yields in arid regions, its causes and possible remedial measures. Hydrol. Symp. Nairobi, pp. 281—289.

9. WATER RESOURCES DEVELOPMENT

9.1 GENERAL

Because seasonal alternations of surpluses and deficits of water cause a basic problem in the utilization of African water resources, three main problems are encountered in water-resource development schemes:

a) Relation of the water resources to the environment,

b) the search for the most rational way of using water resources,

c) the search for the most rational manner of their development.

When solving them, it must be borne in mind that the conditions of the tropical resources are different from those in moderate regions and any experience from outside the tropics should be applied with caution. The different conditions are characterized by a higher rate of energy flow through the tropical ecosystems, which is caused by higher temperature and greater potential productivity. This leads to a higher consumption of nutritive salts and a greater need for their replenishment. Waste products and dead organisms rot more rapidly and this requires a higher consumption of oxygen. There is also a higher evaporation and evapotranspiration rate in the tropics.

As a continent, Africa receives 670 mm of rainfall in a mean year, 510 mm of which evaporates and 160 mm flows to the ocean. The mean annual discharge is estimated at 105.4×10^3 m^3/s flowing from an area of $18,700 \times 10^3$ km^2. The lakes of Africa contain 36 000 km^3 of water, which represents about 30% of the world freshwater total and, according to some ecologists, the world of tropical Africa is dominated by its lakes, not by its rivers. The total suspended-sediment yield from the continent is estimated at 1395×10^3 tons per year, which is a surprisingly low value when compared with other continents. Similarly, the dissolved matter yield of 757×10^3 tons per year is also low.

48% of the continent is drained into the ocean, 40% is without surface drainage and 12% is drained into interior basins.

9.2. DATA NEEDED FOR DEVELOPMENT

Data on various aspects of water resources are essential for water-resource planning and implementation in the tropical regions, because any project in water-resource management, in the African tropical environment, must be viewed from

182

the aspect of its basic limitations and any attempts to overstep the limits may result in great failures. In addition to adequate hydrological data, the need for a reliable inventory of available water resources in the tropics has been stressed several times at meetings of the Economic Commission for Africa. The situation calls for the establishment of regional or continental data banks. Beside hydrological data and analytical bibliographies, economic and sociological information related to the water should be established as the basis for assessing the various aspects of water-resources projects and for comparing water requirements and available water resources.

The need has been stressed on many occasions for all African countries to have their meteorological and hydrological services organized in an integrated form and therefore adequate communications must be established to ensure rapid transmission of observed data. Though much has been done to solve this problem by the World

Tab. 9.1. Network density as recommended by WMO for tropical regions

Region	Minimum network	Provisional network
	Area in km^2 for 1 station	
Principitation stations:		
Flat	600—900	900—3000
Mountainous	100—250	250—1000
Mountainous islands	25	
Arid	1500—10,000	
Evaporation stations:		
Humid	50,000	
Arid	30,000	
Stream gauging stations:		
Flat	1000—2500	3000—10,000
Mountainous	300—1000	1000—5000
Mountainous islands	140—300	
Arid	5000—20,000	

Meteorological Organization, unification is still far from complete. The hydrological and meteorological services must be extended in many countries before reasonable data for water-resource planning and design will be available. International standards should be adopted in the collection and analysis of basic data. Tab. 9.1 gives the standards for network density as recommended for Africa by the WMO.

Data should be made available not only for large schemes, but also for the large number of small utilizers. They should also be published in a form which permits easy orientation and comparison with the data from neighbouring countries.

In principle, the network of stations should cover typical tropical features, such as convective vertical motions, banded vertical motions, general vertical motions associated with large synoptic features and, last but not least, the topography of the region. In representative and experimental catchments, an additional network should cover the spatial and time variability of rainfall which is typical for tropical Africa. The catchments should be selected in areas which are in many ways typical tropical ecosystems.

An evaporation network should preferably be established in moisture deficiency zones, irrigated areas and near completed or planned reservoirs. Meteorological stations, equipped for energy budgeting, should be established nearby.

The selection of streamflow and water-level gauges should be made with a view to the possibility of interpolation of the data and, before some recognized standard is achieved, attention should be paid to the stations on large rivers and locations of economic importance. If possible, streams with natural regimes should be gauged.

Water-quality measurements should be taken in streams in areas with high population or industry concentration. Beside standard chemical analyses, special bacteriological/biological observations should be organized, spread over the rivers and lakes for the control of tropical diseases.

Groundwater observation should be organized in a way that will allow an estimation to be made of the ability of an aquifer to yield water. Therefore, the observing system need not necessarily be distributed uniformly throughout the region, but in keeping with its geological structure. An additional system of boreholes and wells should be installed for special purposes, in representative areas or mining regions, for instance.

9.3 INTERNATIONAL DRAINAGE BASINS

The tropical part of the African continent has forty-eight international drainage basins. About 40% of the whole continent can be defined as an international basin. While earlier treaties on African international rivers were concerned with the navigation and boundaries, more typical of recent treaties is the emphasis on economic development. Since lack of data is one of the first difficulties met with in this type of project, collaboration among the states in the initial stage is in the collection and exchange of data. Thus, international cooperation quite literally starts with hydro-meteorology.

Table 9.2, provides information on the international basins of tropical Africa. The table is based on surface runoff boundaries. Groundwater flow may be found to be an additional complicating factor.

Tab. 9.2. International rivers in tropical Africa

No.	Name of river	Countries	Approximate basin surface area (km²)
1	Senegal	Guinea, Mali, Mauritania, Senegal	441,000
2	Gambia	Gambia, Guinea, Senegal	77,000
3	Geba (also called Kayanga)	Guinea, Guinea Bissau, Senegal	8,000
4	Corubal	Guinea Bissau, Guinea	20,000
5	Kolenta (Great Scarcies)	Guinea, Sierra Leone	8,000
6	Little Scarcies	Guinea, Sierra Leone	15,300
7	Moa	Guinea, Liberia, Sierra Leone	17,900
8	Mano Roro	Liberia, Sierra Leone	9,000
9	Loffa	Guinea, Liberia	8,700
10	St. Paul	Guinea, Liberia	18,300
11	St. John	Guinea, Liberia	15,000
12	Cestos	Ivory Coast, Liberia	10,300
13	Cavally	Guinea, Ivory Coast, Liberia	22,400
14	Bia	Ghana, Ivory Coast	9,320
15	Tano	Ivory Coast, Ghana	15,000
16	Volta	Ivory Coast, Ghana, Mali, Upper Volta, Dahomey	390,000
17	Mono	Dahomey, Togo	22,000
18	Oueme	Dahomey, Nigeria	50,000
19	Niger	Cameroon, Chad, Dahomey, Guinea, Ivory Coast, Mali, Niger, Nigeria, Upper Volta	1100,000
20	Cross River	Cameroon, Nigeria	48,000
21	Chad	Cameroon, Chad, Central African Republic, Niger, Nigeria	—
22	Ntem	Cameroon, Gabon, Rio Muni	31,000
23	Benito	Gabon, Rio Muni	14,000
24	Utambori (Rio Temboni)	Gabon, Rio Muni	5,000
25	Ogooue	Cameroon, Congo (B) Gabon	205,000
26	Nyanga	Congo (B), Gabon	18,000
27	Chiloango	Cabinda Congo (B)	13,000
28	Congo	Angola, Burundi, Cameroon, Central African Republic, Congo (B), Zaire, Zambia, Tanzania, Rwanda	3820,000
29	Cunene	Angola, South West Africa	100,000
30	Cuvelai-Etosha	Angola, South West Africa	—
31	Okavango	Angola, Bechuanaland, South West Africa	—
32	Orange	Basutoland, South Africa, Bechuanaland, South West Africa	640,000

Tab. 9. 2. (Continued)

33	Maputo	Mozambique, South Africa, Swaziland	35,000
34	Imbuluzi	Mozambique, South Africa, Swaziland	6,400
35	Incomati	Mozambique, South Africa, Swaziland	30,000
36	Limpopo	Bechuanaland, Mozambique, Rhodesia, South Africa	358,000
37	Sabe (Save)	Mozambique, Rhodesia	101,000
38	Buzi	Mozambique, Rhodesia	32,000
39	Zambezi	Angola, Bechuanaland, Malawi, Mozambique, Rhodesia, Zambia	1250,000
40	Ruvuma	Mozambique, Tanzania	140,000
41	Lake Natron	Kenya, Tanzania	—
42	Lagh Bor	Ethiopia, Kenya, Somalia	—
43	Lake Rudolf	Ethiopia, Kenya, Sudan	—
44	Juba	Ethiopia, Kenya, Somalia	200,000
45	Shebile (Wabi Shebele)	Ethiopia, Somalia	260,000
46	Gash	Ethiopia, Sudan	21,000
47	Baraka	Ethiopia, Sudan	—
48	Nile	Zaire, Ethiopia, Kenya, Sudan, Tanzania, UAR, Uganda, Rwanda, Burundi	2800,000

International agreements on African rivers can be traced back as far as 1885 when navigation of the Congo and Niger for ships of all nations was announced at the Berlin Conference. A similar treaty between Great Britain and Portugal covered navigation for all nations of the Zambezi, Shire and all branches including landways at parts where the streams were not navigable. An agreement on River Nile management was signed in 1929 between Great Britain and the Egyptian Government. A similar agreement on technical cooperation in basin development was signed between the Sudan and the United Arab Republic in 1959. The Senegal basin development scheme was announced by four states sharing the basin in 1969 and an Interstate Committee is seeking ways of developing the whole basin in an integrated way. Two Acts on the Niger basin were signed in 1963 and 1964 which abrogated former agreements and declared its complex utilization. The Chad Basin Commission was established by a special convention in 1964.

There are still some controversies on the international rivers of tropical Africa, such as the point on the Zambezi where the boundaries of South Africa, Southern Rhodesia, Zambia and Botswana meet. The boundaries are partly defined here as a thalweg and partly as a line in the middle of the river and they do not coincide at the critical point.

In general, agreements based on mutual understanding of national interests can help to further development of the vast tropical water resources and facilitate the financing of surveys and water works and the exchange of hydrological information. A great number of projects can be realized in cooperation with United Nations

Organization which assists during the preparation and initial stage of programs. A typical example of such a project is the hydrometeorological survey which was organized by WMO on Lakes Victoria, Kyoga and Albert and with the cooperation of Uganda, Tanzania, Kenya, Sudan, UAR, Rwanda and Burundi.

9.4 NAVIGABILITY

The conditions for river transport are not very favourable in tropical Africa. The Atlantic side of Africa has more navigable rivers then the Indian Ocean side. Between the rivers Senegal and Cunene there are 36 rivers that are navigable in their lower courses by launches or by bigger ships for at least part of the year. Between Juba and Incomati, there are only seven. Most of the navigable rivers are short and with ungraded courses. Owing to the heavy surf and shifting sands, the connection between oceans and the interior is very limited. The Senegal is an exception and is navigable for 800 kilometres, at least for small launches. The Saloum and Gambia are other significant waterways connected by their distributaries. The rivers Casamance, Cacheu, Geba, Nunez, Pongo, Mellacorée, and Sierra Leone are the most significant waterways north of the equator. The last of these, together with its tributaries, deltas and branches, has 800 kilometres of routes.

Because of the heavy surf produced by the Guinea current, many rivers flowing into the Guinea Gulf are cut by sandbars. There is launch traffic on the Tano, Ankobra and Lower Volta, but no seagoing traffic. The rivers Mono in Togo and Ouémé in Dahomey carry small launches for about 150 kilometres into the interior.

The mainstream of the Niger is navigable to Jebba; however, its navigability greatly depends on the season of the year. The Benue is navigable by shallow-draft vessels. The gross river system east of the Niger can transport small vessels for almost five hundred kilometres upstream. The rivers Mungo, Vouri and Dilamba are navigable for 80 kilometres into the interior and, at high water, small launches can pass to the Sanaga river, which is navigable in its lower reaches by small craft. The Nyong, Muni and Gabon are navigable for 160 kilometres and one can reach the system of the Ogoué by small boat. This river is navigable up to Lambaréné. Between the Ogoué and the Congo, the river Koilou is navigable for only 130 kilometres. The Congo itself is navigable from the ocean for about 130 kilometres. With the exception of the Cuanza, the streams south of the Congo are navigable for a few weeks per year and the Cuanza is navigable for 200 kilometres in a good year.

On the Indian Ocean side, the streams Incomati, Limpopo, Macuse, and Zambezi are capable of handling some traffic. The Zambezi is navigable up to Tete and Malawi can be reached by small vessels through the Shire. The rivers Ruvuma and Rufiji can take small vessels. The River Tana is navigable up to 600 kilometres by small vessels in a good season. The Juba can be travelled by small boats only for 650 kilometres between May and December.

Perhaps the inland waterways are more useful. The Congo system has 13,000 kilometres of navigable waters and almost 3500 of them can be travelled by large barges. The middle Niger is divided by the rapids at Bamako and Koulikoro into two sections, while the navigable section of the main Congo is divided into five parts. There is very little transport on the Nile river because of six cataracts in its course. Traffic is limited to the sections between them. Three hundred km of the Blue Nile are also navigable. The White Nile is navigable for 150 kilometres above Khartoum. Upstream the Juba has rapids up to the Uganda borders, from here the river is navigable to Lake Albert. The Sobat and Bahr el Ghazal tributaries of the Nile are also navigable.

All great African lakes have regular water transport, the small ones and the swamps only being suitable for small boats.

9.5 IRRIGATION AND FISHERIES

Food potential dependent on water, is still very much underdeveloped. Although the total food supply in the region is adequate, there is a serious nutritional imbalance and quantitative seasonal deficiencies in some regions, according to UN sources. Although about 82% of the population is enagaged in agriculture as the primary occupation, the area presently under cultivation is only 10% of the total land. Rain-fed agriculture is the most practised. In the near future, the cultivated areas can be doubled, at least.

The extent of irrigated agriculture is closely related to the ecological zones and its intensification is recommended in:

a) arid lands where the settled agriculture depends on irrigation,

b) semiarid areas to ensure safe yields,

c) in areas with erratic rainfall for the production of high-value export crops,

d) in areas suitable for crops with high water requirements.

The first type of land is the most difficult to irrigate because the climate is too dry to permit successful growth of crops in average years. Usually, areas in the outer margins of semiarid regions; semiarid to arid areas and areas of extreme aridity are recognized. In the outer margins, there is no risk of inadequate rainfall, in the second, there is a chance of cropping in some years, in the third, there are twelve months without any rainfall. The African deserts Kalahari, Sahara and Somali Chalbi are the most extensive. The Kalahari receives less than 50 mm of rainfall per year, but benefits from sea fog and dew which support sparse vegetation and ephemeral grasses. In Northern Kalahari, neither the climate nor the vegetation have the character of a true desert. The lack of water is caused by deep sand layers. Authors differ in their assessment of Namib part of Kalahari between extreme desert and desert. The Sahara is the largest desert in the world and a great part of it has less than 25 mm of rainfall per year. The vegetation cover is sparse, susceptibility to wind erosion is very light, infiltration rates good. The Somali Chalbi is a narrow band along the Red Sea between the Gulf of Aden and the Indian Ocean, south of the equator. The low rainfall is related to orography. Usually, less then 100 mm are concentrated in five months. Grasses and undershrubs are scattered over the area while shrub and tree savanna start in the margins having 200 mm of rainfall.

If for no other purposes, water is needed in the arid and semiarid areas for nomadic herding, for hunting and fishing, for stock raising and for agriculture. Where emphasis is placed on irrigation, schemes should involve other needs as well.

Water supply estimate for the irrigation schemes is usually provided first. Areas are recognized where

a) water is not available at all,

b) water is available in small local supplies such as pan valleys,

c) water is available only from intermittent streams,

d) water is available from areas with humid climate,

e) water is available from groundwater sources,

f) water is available from only occasionally replenished streams.

It is surprising that some areas that are now arid or semiarid once used to be naturally fertile and were exploited for extensive agriculture. This means that the ecological conditions have changed due to large-scale fluctuation in the general circulation. This should be considered as a potential danger elsewhere in the margins of arid and semiarid regions.

The situation may be improved in some places simply by removing the pheato-phytes and upland scrub and thus reducing evapotranspiration. Frequently, more expensive arrangements are necessary, such as the conservation of flash floods in small reservoirs or their direction to the fields or to the plains with the aim of recharging the aquifers. Wind-breaks could be another type of reclamation accompanying irrigation schemes. During the preparatory stage, the possibility of the quick spread of pests and diseases related to water should be always considered.

It has been concluded from economic analysis, that priority should be given in the African tropics to small-scale village-type irrigation schemes which can be easily managed and financed and in which farmers can fully participate. Where a big scheme might be a failure, small projects have a limited loss. The FAO recommends medium-sized irrigation schemes on 100—10,000 acres in addition to small ones. Big projects of the type of the diversion of the River Niger into the interior of the Sahara have been proved risky. In Ghana, for instance, it was concluded that before any large-scale development is undertaken, all efforts must be directed to the improvement of small garden-type irrigation schemes in the tributary valleys. Similarly, the shores of Lake Victoria in Kenya have been surveyed for irrigation and about 14,000 hectares selected. 800 hectares were chosen for the establishment of the pilot scheme including an irrigation research station. A similar project has been initiated in the Kafue Flats of Zambia.

It is foreseen that almost 2000,000 hectares will be irrigated in Africa and Malgasy by the late nineteen-eighties. The present situation is shown in Tab. 9.3, indicating arable land, harvested land and irrigated areas.

Tab. 9.3. Arable, harvested and irrigated land in million hectares

	Arable	Harvested	Irrigated
West Africa	75	35	0.2
Central Africa	25	6	0.01
East Africa	52	23	0.9

The role of the hydrologist can be seen in the preparation of adequate data, particularly for small streams which may be considered convenient for an irrigation water supply. Apart from this, studies concerned with large-scale projects and proposed climitical changes should be carefully prepared.

Drainage is a problem of equal importance in some regions. Agricultural land is frequently damaged by excess water and by its removal. Flooding, waterlogging and increased salinity are the main problems which can usually be solved by drainage and flood-control techniques. In the Yala basin, for instance, the diversion of the river protects some 20,000 hectares of fertile soil from flooding. Regulation of the Shabelli river in Somaliland was also necessary before any controlled irrigation could be provided. Subsurface irrigation has not been widely used in the tropics.

Listed in Tab. 9.4 are some of the largest irrigation schemes in Africa.

Tab. 9.4. Some of the large irrigation schemes in tropical Africa

Country	Project	Purpose
UAR	Dams and drainage	108,000 acres at Kom Obo
Upper Volta	Planned agric. dams	at Kouougu
Ghana	Accra Plains Project	440,000 acres in Nothern Region, cotton and tobacco
Nigeria	Water supply scheme	Ibadan area
Burundi	Water supply scheme	Bujumbura area
Kenya	Dams and drainage	Kano Plains, Lake Victoria
Botswana	Dam and drainage	Sashi river project
Senegal	Gunia dam	Senegal river scheme, interstate project
S. W. Africa	Cunene river scheme	power and irrigation for Owambo
Zambia	Kafue Flats	Irrigation scheme

Fisheries are an equally important source of food from African waters. The rivers and lakes are very rich in variety of species. Lake Nyasa, for instance, has 200 species of fish which is the same as in the coastal waters of the Indian and Atlantic Oceans. However, there is very little commercial interest in the fish industry as yet because of transport difficulties. For example, the dried fish, kapenta, is supplied only to local markets in Zambia which are a short distance from the fishing centers.

Although fish is not looked upon as a staple diet by many tribes, practically all the rivers are fished at some time or other; only the small torrential streams are not suitable for fishing. It was proved recently that the annual commercial production of fish in small ponds or intermittent streams carrying water for about 9 months could be 8 – 10 times higher than in Europe.

The fishing techniques are closely related to the hydrological regimes of the rivers. During the rising periods of the Zambezi and its tributaries, the Lozi tribe uses no-return traps. When the river falls, fish dams and fences are preferred. In the swampy areas along the rivers, the fishermen are semi-nomadic. moving their settlements so that they remain in close contact with the river, as it rises or falls. Fishermen in the Lukanga swamp live on floating islands inside the swamps. Lake fishing has an even higher potential than that of the rivers. In artifical lakes, however, the total production is much smaller. In Kariba, for instance, the total catch was six times smaller after two years. Man-made lakes are thought to be less advantageous for the fishing industry when left under natural conditions. The situation requires the establishment of artifical pilot fishing schemes and permanent artifical improvement of introduced species. Very likely, the establishment of small shallow ponds will prove to be more effective than the utilization of large dams, providing the hydrological conditions are favourable. In some localities, a combination of irrigation schemes with fishing industries may be found to be economical.

Again the role of the hydrologist is important, particularly at the preliminary stage of the project planning, in searching for the best locations for ponds and water supplies and in determining the duration and seasonal distribution of flow.

9.6 RURAL WATER SUPPLY AND WATER CONSERVATION

In general, two ways of increasing the water supply are recognized in arid and semiarid regions:

a) Through improvement of the existing hydrometeorologic conditions,

b) by improving the possibilities of making full use of the existing water supply potential.

The list of sources which can be utilized in the tropics includes:

a) Surface streams which maintain water all the year,

b) subsurface streams found at locations with gentle slopes, where the water moves slowly into the stream. (At some Sahara localities up to 100 l/sec. is gained from those sources),

c) phreatic water which is normally thought to be of poor quality (saline), but which is easily obtainable from shallow wells and boreholes in coastal dunes,

d) artesian water that can be pumped or which flows up to the surface,

e) springs, which frequently supply only a limited amount,

f) artifical tanks collecting torrential water.

In general, it must be taken into consideration that some sources supply water at the expense of other regions and frequently it is water which cannot be replenished during our era. There are six regions in the tropics which differ in water supply potential:

a) Extremely arid regions with several years without rainfall. The mean potential supply will be 50−100 mm per year and a rather insignificant increase is possible only through long-distance transport of water.

b) Sparse localized supplies are found in the areas with 100−400 mm of potential rainfall. Some water can be collected in localized areas. Surface flow is ephemeral, erratic and violent in form. The water seeps from stream channels into valley sediments and evaporates in shallow playas. Small basins of groundwater supply and river beds are replenished and water is locally available. Desert areas near the Nile are typical examples of such a region.

c) Semiarid supplies, found in the regions where the seasonal flow in the rivers varies from year to year, but where some perennial streams are available as well as underground sources. Sufficient precipitation supports the crops in some years and there is a possibility of improving the supply by the construction of small dams and by tapping shallow wells.

d) Exogenous surface water supplies are found at locations where extremely arid areas are crossed by major rivers with a perennial flow forming a high-potential groundwater supply. Areas along the river Nile belong to this type of region.

e) Exogenous groundwater supply regions are in arid or semiarid areas where the underlying aquifers are replenished from humid or semihumid areas.

f) Localized exogenous supplies are found in the areas where a sharp difference in relief and rainfall exist over a short distance. Arid valleys are fed by small aquifers from mountain chains in their vicinity.

g) Endogenous supplies are found in the areas where the local supply potential exceeds the actual evapotranspiration and seepage and water supply is in excess of the need of crops and human beings.

The criteria for local supply potential are very variable. For instance, a Bushman in Kalahari needs only 1 l/day, as compared with an inhabitant of a well-developed industrial region who may need up to 800 l/day.

Rural water schemes are normally not self-supporting and have to be built with governmental assistance. Even non-expensive schemes can impove living and healht conditions significantly. The WHO reported that health conditions definitely improved through the rural supply system in the Zaina area, Kenya, to 5800 persons. This was proved by comparison with a neighbouring non-supplied area. The incidence of some diseases was six times lower and economic development was faster.

In Gwembe valley, Zambia, an experimental water collecting tank was built on the left bank of the Zambezi in the area where the Ila-Tonga people suffer from a periodic water shortage. The tank was built from locally available materials and protected against evaporation by a grass cover, as a self-help scheme. It is now used by the people living within distances of up to 8 kilometers and the need for another tank is urgent. Simple materials, such as locally made bricks or sand sausages (sand filled plastic bags) and sorghum grass make this sort of construction very cheap.

An extensive programm for the conservation of water should accompany any long-term planning of water resources utilization. Careful application of farming practices, soil fixation through rapidly growing plants, terracing, contour cultivation, grazing control, afforestation, etc., are simple, inexpensive means which should be developed and supported by the administrative organs. It has been proved experimentally in East Africa that proper fertilization of the soil can reduce runoff and erosion significantly.

Systematic improvement in appraisal of water resources, reconnaisance surveys, intensive observations in experimental and representative basins selected in ecologically different regions, systematic silt sampling and measurement of the sediment load of the rivers are the main tasks of hydrologists in the field of water resources development.

9.7 DAMS AND LARGE SCHEMES

In one of its stretches the Congo river descends almost 300 meters through 32 falls and cataracts over a distance of 350 kilometres. The river develops the potential of 114 million horsepower. The potential of the whole Congo basin is 168 million horsepower, about 15—25% of the world potential. Although this river is exceptional in many ways, there are other large, medium and small streams which, together with the Congo, develop a potential of 260 millions horsepower, or in other words, 37% of world total. The parts of Africa with 1500 mm of rainfall and which are more than 300 meters above sea level, are the most favourable for major schemes.

In historical development, the large projects have been undertaken in Africa not only for power, but as multi-purpose schemes. In addition to the historical schemes, the first modern one was the Gesira project developed from 1913 up to 1950. It consists of a system of irrigation canals and the Sennar Dam which directs water from the Blue Nile into the main canal. Crop rotation includes cotton, millet or sorghum, beans, cotton and fallow periods. The scheme, with the Manaqil extensions receiving water from the Roseries Dam, covers over 728,000 ha.

The largest man-made lake was created by the construction of the Kariba Dam on the boundaries of Zambia and Southern Rhodesia. The postwar growth in industry and agriculture on both sides of the Zambezi resulted in a great demand for energy. In the nineteen fifties the increase of energy consumption was 11% per annum. The project for a large dam was found at that time to be the most economical solution, compared with the alternative of a nuclear station or a number of smaller power stations. The project also ensured some control over the Zambezi floods and some extension of shipping services. The dam produces 600 MW of electricity, which is split between Zambia and Rhodesia. In the seventies, the Zambezi left-bank scheme was initiated in an attempt to increase power for Zambian industry. The Kafue scheme on the tributary of the Zambezi, which was finished in the seventies, provides another 900 MW, although the river is far smaller than the Zambezi.

The Aswan Dam, although outside the tropics, should be listed in the African tropical schemes, because the reservoir reaches deep into the tropics. The total capacity of 2100 MW and perennial irrigation and land reclamation of the lower Nile are the main economic results. The old Aswan Dam had twice been elevated to improve its water capacity, but the amount of silt accumulated in the old reservoir prevented storage of sufficient water. The new dam increased the total area of irrigated land from 4% to 25% of the whole country.

Another of the large African schemes was initiated in Ghana 1962. The long, dry season on the savanna river Volta and shortage of energy for the mining industry were the main reasons for the development of the scheme. The establishment of secondary industries was equally important. The lake behind the Akasombo dam covers 3% of Ghana's surface. The dam also provides water for irrigation of the Accra plain and the area north of the lake. An inland waterway was built up to the lake and the installed capacity was 768 MW.

The Kainji Dam on the Niger in Nigeria provides some 320 MW and will be extended to 960 MW. Some low-cost potential sites on the tributaries and on the Niger itself have already been chosen for further development.

A comparison of four major dams completed in Africa is given in Tab. 9.5.

Tab. 9.5. Comparison of the largest African dams

Reservoir	Height m	Area km^2	Altitud. m.a.s.l.	Mean depth m	Ratio of annual outflow to volume	Number of people resettled
Kariba	124	4300	530	125	1 : 9	50,000
Volta	113	8500	92	70	1 : 4	70,000
Kainji	145	1280	155	55	4 : 1	50,000
Aswan	111	5000	185	97		120,000

Beside the big schemes, a great number of smaller dams are being planned or completed. The Limpopo floods are controlled by a dam 70 meters high, which also provides water for irrigation. The Owen Falls dam, at the Nile outflow, was built in 1954, is 26 meters high and produces 105 MW, which is about 15% of Uganda's present needs. The Coka dam was built on the Awash river some 80 kilometres from Addis Abeba with an installed capacity of 46 MW. A large dam in Mozambique, the Caborra-Bassa, will produce in its final stage 1200 MW, flooding 365 km^2, and a dam on the Mono river between Togo and Dahomey; the Kossou dam on the Bandama river (174 MW) in the Ivory Coast and the Konkouré Project (1200 MW) in Guinea are being planned. The 300 MW Goina Dam is being planned in Senegal, Mali.

In recent years, a great number of super-large schemes has been proposed by various authors. Sergel recommended flooding about 10% of Africa by forming a so-called Congo Sea, high enough to divert the Congo waters into the Lake Chad basin and irrigate part of the Sahara. In the Kalahari, the restoration of the inland lake was recommended by the weiring of the Cunene and Chobe rivers, giving an expected increase in rainfall of 250 mm. The general scheme of an Upper Nile Project was based on the century-storage lakes and attempted to utilise fully the clean waters of the Upper Nile reservoir and to project the water against evaporation in the vast swamps of Southern Sudan. At the beginning of this century, a project for the diversion of the Niger waters into the central Sahara region was proposed and, later on, a special Niger Office was established, under which the Diamarabougou dam was built with the aim of reirrigating the land once under flood from the old channels of the Niger which used to flow into a vast lake at Timbuktu. Apart from technical difficulties, the population was not prepared for such a large resettlement.

Actually, problems of resettlement almost always arise when a large scheme is under preparation. Resettlement projects, which should be prepared well in advance before the constructional works start, are realized at the last moment and without giving much thought to sociological aspects. Thus, people are shifted to a strange or even hostile country, where any adaptation to existing conditions is difficult if not impossible. This happened during the construction of the Kariba dam to the tribes which had been living in the fertile river valley for hundreds of years. In many cases, the rural population leaves for the towns and becomes less progressive and dependent on government welfare for indeterminate period. In some areas, a 10% increase in the death rate was reported. In Rhodesia, famine maize relief continued for more than twenty years.

Beside resettlement problems, there are other negative aspects of the large schemes. The constructions, particularly in the Rift valley, are subject to cause earthquakes, because of the enormous weight of the accumulated water. Another problem is the control of weeds and aquatic vegetation, particularly in the initial stage of the lake development. A similar problem is the spread of tropical diseases such as bilharzia.

The deposition of silt in the reservoirs, which drastically reduces their efficiency and which has been experienced particularly in the Nile region, is a problem still to be solved.

For these reasons, other power sources, such as solar energy, should be examined. It has been calculated that in tropical areas with bright sunshine, a power station with an output of 50 MW can be constructed on an area of 2.5 km^2.

Future development in Africa will most likely be concerned with small and medium-sized schemes, rather than with large ones. The first give greater effect in a shorter time and are easier to manage and control with local manpower.

9.8 ENVIRONMENT AND POLLUTION

The worldwide problem of environmental protection is not yet fully developed in Africa, because the degree of industrialization is still small. Main problems are still seen as nature preservation, game protection and protection of the ecosystems, and environmentalists are frequently opposing any possible threats to African nature by dams, irrigation schemes and agricultural development. On the other hand, environmental damage may be caused by unreasonable conservation policies. At one time the Luangwa game reserve became so overherded by elephants, that wide areas were seriously damaged, losing their vegetation cover and the soil was exposed to intensive erosion. This, of course, negatively influenced the hydrological cycle. Problems of such a type usually require complex solutions based on sound inter-disciplinary research and hydrological research alone is of limited use.

Artifical pollution of natural water resources in the tropics is still rare and is limited to the vicinity of big towns and mining centers, although occasionally we may witness a tropical stream polluted near villages by the foam of detergents. A much greater hazard exists in biological pollutants, found traditionally in tropical streams and, in some cases, artifically induced. The quality of tropical surface water must always be regarded with suspicion, particularly at the end of a dry season, when concentrations are much higher. If regular water treatment is not possible, sand filters, tubewells in the banks, or infiltration galleries at least should be constructed. Many surface waters, though harmless to the local population, may be a source of disaster for people from outside the region.

Groundwater sources, except for shallow wells, which however, are traditionally avoided by many tribes, are much safer. Chlorination is a simple method, but not a substitute for filtration.

In slowly-moving shallow waters, vegetational pollution is a serious problem. The water hyacinth (*Eichornia crassipes*) is widespread in the Nile basin and many methods including spraying from aircraft, have been tested to keep it under control. Snails hosting the fluke which causes bilharzia, feed on this type of tropical vegetation, particularly along the banks and in shallow waters. The disease is endemic to the tropics, and is second only to malaria, which is also related to standing water series. In some areas with developed irrigation systems, bilharzia affects 75% of the population. Apart from costly prevention measures, wide-scale health education is one of the most effective ways of combating diseases.

It can be concluded that tropical resources may be reasonably developed without drastic deterioration of the environment, providing the most basic protective measures are included in the projects at the initial stage of planning.

9.9 WATER POLICY

Water institutions, administrations and legislations in tropical Africa are derived from several legal systems. Chief among them are African customary law, Moslem law and Roman law, the latter being adapted to suit the needs of the previous colonial systems.

African customary law varies from country to country. It covers land ownership, cultivation, watering and fishing rights. According to this law, private ownership of water was generally unknown and individuals had the right to use water, while land tenure was communal or tribal. In the future, this may simplify various problems of centralization of state water control.

According to Moslem law, water entails a religious obligation derived from its nature and no one can refuse surplus water without sinning against Allah and man. All members of a Moslem community are ensured water availability. According to the Moslem traditions, all waters are deemed to be common property. Since the principles of Islamic law are strictly followed in tropical Moslem countries, any modernisation of water administration should take them into account.

Roman law divided water into three categories, private, common and public. Private water, springing or flowing within one's property, is part of the property, common is any unowned water, such as rain, the seashore and non-navigable rivers. Public water includes large rivers, canals, lakes, which are generally navigable, and is governed by public law.

A combination of the three laws makes the existing situation in water-resource administration rather complicated, particularly with a view to the rapid political changes on the continent. It is foreseen that all water will be brought under government control to ensure its efficient utilisation.

9.10 WATER-RESOURCES PLANNING

In the planning of water-resource development, an integrated point of view should be emphasized. Planning programs should be flexible, with priorities emphasized and should contain an appraisal of alternative solutions to the problems presented. According to past experience, priority should be given in the planning of tropical water resources to small and medium sized development schemes and to projects in which implementation is most likely to follow immediately after their completion. Self-help programs with some supplementary assistance from government agencies as potential supporting sources should be considered. International assistance in terms of experts, salaries, equipment and supplies, should be considered as marginal. Each plan should be based on adequate data and if the data are not available, a solid network should be established within the area of the scheme. Basic and applied research should be connected with the problems of existing and planned schemes and with the problems of more extensive and complex schemes. Problems of man-power should also be solved at an early stage at all levels, from men in the field to postgraduates. Training programs should in fact become a part of each development plan.

Problems of inter-African cooperation should be solved, where by the schemes could benefit, directly or indirectly, other neighbouring states. A comparison between available water resources and present and future needs should become a part of the plan, particularly in regions where the already existing water surplus is low. The relationship between the water economy and the national development plan should always be carefully examined.

9.11 LIST OF LITERATURE

[1] Economic and Social Council of United Nations reports.
[2] Economic Commission for Africa reports.
[3] FAO reports and citations.
[4] McGinnies, W., G., Goldman, B., J., Paylore, P., 1970. Deserts of the world. Univ. of Arizona Press.
[5] Kimble, G., T., 1960. Tropical Africa. The Twentieth Cent. Fund.
[6] Mackenzie, C., A., 1945. Reports on the Kalahari Expeditions. S. African Government Printer.
[7] Meigs, P., 1966. Geography of coastal deserts. UNESCO Press, Arid zone research 28, 140 p.
[8] Sinaika, Y., M., 1964. Alternative uses of limited water supplies in the Egyptian region of the United Arab Republic. The problems of arid zones, UNESCO, Paris.
[9] Uvarov, B., P., 1962. Development of arid lands. The problems of arid zones, UNESCO, Paris.
[10] White, G., F., 1967. Alternative uses of limited water supplies. The problems of arid zones, UNESCO, Paris.
[11] WHO reports and citations.
[12] WMO reports and citations.

APPENDIX 1. FLOW CHART OF THE TROPICAL WATER BALANCE MODEL

see enclosure

APPENDIX 2. FORTRAN SYMBOLS IN THE MODEL

1, 2	is the balance value at the beginning and end of the balance interval
AA	Drainage area, m^2
AF	Coefficient of the evaporation equation
AI	Current infiltration rate into the surface biological zone, in/hour
AIXX	Minimum infiltration rate into the biological zone in/hour
AIYY	Maximum infiltration rate into the surface-biological zone, in/hour
BX	Calendar number of a day when the leaves are most developed
BB	Calendar number of a day when the roots are most developed
BF	Coefficient of the evaporation equation
DIF(A)	
DIF(B)	Coefficients of the differential equation
DIF(C)	
ED	Potential evaporation, in
FDY	First day of the balance calculation
FFF	Rate of capillary rise, in/hour
FFX	Rate of percolation, in/hour
FI	Instant infiltration rate into the soil, in/hour
FMX	Infiltration into the soil as released by the surface-biological zone (in/hour)
FPOR	Porosity corresponding to field capacity (in/hour)
GWX	Groundwater storage in the groundwater zone, in
HC	Groundwater level calculated as a depth below surface, in
(I)	Symbol for a region
(J)	Hourly index
K	Recession coefficient
LMA	Storage in upper moisture zone, in
LMB	Storage in lower moisture zone, in
LME	Maximum capacity of the lower moisture zone which can be reached by the capillary rise, in
NPOR	Noncapillary porosity, in
P	Precipitation, in
POR	Capillary porosity, in
QDA	Outflow from the deep region, cu.sec
QO	Discharge observed, cu.sec
QRA	Surface runoff outflow, cu.sec
QSA	Outflow from shallow region, cu.sec

RH	Instant lenght of the roots, in
RHXX	Min. depth of the roots in a year, in
RHYY	Max. depth of the roots in a year, in
SG	Maximum capacity of the groundwater zone, in
SL	Total capacity of both soil moisture zones, in
SLA	Storage in upper moisture zone, in
SLB	Maximum capacity of the lower moisture zone, in
SU	Current capacity of the surface-biological zone
UM	Storage in surface biological zone, in
WPOR	Porosity corresponding to the wilting point
XK	Ratio of deep to shallow region
XX	Minimum capacity of the surface-biological zone during a year, in/hour
YY	Maximum capacity of the surface-biological zone during a year, in/hour

APPENDIX 3. LIST OF ORGANIZATIONS CONCERNED WITH AFRICAN TROPICAL HYDROLOGY

For those hydrologists who might be concerned with the utilization of the hydrologic and meteorological data from tropical Africa and with various projects in the international basins, a list of the organizations, institutions and services collecting data and cooperating in the field of tropical hydrology is attached.

With a view to the rapid development of hydrology and related fields, new institutions not included in this list, may appear and others may exist under different names. Therefore, whenever possible, more organizations are given for each country, some of them only partly concerned with hydrology, which can at least supply some additional information on where to ask for data.

International organizations

Comité inter-africain d'études hydrauliques, Ouagadougou, Haute Volta
Economical Commission for Africa, P. O. Box 3001, Addis Ababa, Ethiopia
Food and Agriculture Organization, Via dell Terme di Caracalla, Rome, Italy
International Atomic Energy Agency, Kärntnerring II − 13, Vienna 1, Austria
Lake Chad Basin, B. P. 727, Fort Lamy, Chad
Niger River Commission, Niamey, Niger
ORSTOM, Service hydrologique, 19 rue Eugéné Carriére, Paris 75018, France
Organisation des Etats riverains du Sénégal, OERS, Labé, Guinée
Special Sahalian Office, United Nations, New York, N. Y., 10012
United Nations Development Program Offices in the capitals of developing countries, or in the regional capitals
UNESCO Field Service, Office for Africa, P. O. Box 50592, Nairobi
World Meteorological Organization, Hydrology and Water Resources Dept., P. O. Box 1, 1211 Geneva 20, Switzerland

Water Resources Section, Resources and Transport Div., United Nations, New York, U.S.A.

Angola

Servicio Meteorológico de Angola, Luanda
Servicio de Obras Publicas, Luanda

Botswana

Department of Meteorology, Gaborone
Water Affairs Department, Private Bag 29, Gaborone

Burundi

Chef du Service de la Météorologie, B. P. 694, Bujumbura

Cameroon

Le Directeur de la Météorologie, B. P. 186, Douala

Central African Republic

Chef de la Section d'Hydrologie, Bangui
Direction de la Météorologie, Ministère de transport, Bangui

Chad

Bureau de l'Eau, P. B. 830, Fort Lamy
Centre ORSTOM, Service hydrologique, B. P. 65, Fort Lamy

Congo

Service météorologique, B. P. 218, Brazzaville
Service des Voies Navigables, Brazzaville

Dahomey

Service Hydraulique, B. P. 385, Cotonou

Ethiopia

Climatological Institute, Addis Ababa
National Water Resources Commission, P. O. Box 1006, Addis Ababa

Gabon

Météorologie Nationale Gabonaise, B. P. 377, Libreville

Ghana

Meteorological Services Department, P. O. Box 744, Accra
University of Ghana, Legon
Water Resources Research Unit, National Council for Scientific and Industrial
    Research, P. O. Box M 32, Accra

Guinea

Service Météorologique, Conakry.
Ministère de Dévelopment Economique, Conakry

Ivory Coast

Directeur de l'Hydraulique, B. P. 1356, Abijan

Kenya

Department of Hydrology, P. O. Box 30521, Nairobi
Director of Meteorology, P. O. Box 30005, Nairobi
East African Agricultural and Forestry Research Organization, P. O. Box 30148, Nairobi

Lesotho

Hydrological Survey, Ministry of Public Works, P. O. Box 20, Maseru

Malagasy

Service d'Hydrologie, Tananarive
Service Météorologique Nationale, B. P. 1254, Tananarive

Malawi

Dept. of Agriculture, P. O. Box 303, Zomba
Water Resources Division, P. O. Box 34, Zomba

Mali

Service Hydrologique, Direction générale Hydraulique et Energie, B. P. 66, Bamako
Service Météorologique, B. P. 237, Bamako

Mauritania

ASECNA, Génie Rural, Nouakchott

Mauritius

Central Water Authority, St. Paul Road, Phoenix
Meteorological Service, Villemin St., Bean Bassin

Mozambique

Servicio Meteorologico de Mozambique, C. P. 256, Lourenco Marques
Servicio de Obras Publicas, Lourenco Marques

Niger

Service Météorologique, B. P. 218, Niamey

Nigeria

Federal Department of Agriculture, P. M. B. 2164, Kaduma
Hydrological Technical Committee, Meteorological Headquarters, Lagos
Nigerian Meteorological Service, Lagos
University of Lagos, Civil Eng. Dept., Lagos

Rhodesia

Direction of Meteorological Services, P. O. Box 8066, Causeway, Salisbury
Ministry of Water Development, P. O. Box 8132, Causeway, Salisbury
University of Rhodesia, Dept. of Geography, P. O. Box 167 Mount Pleasant, Salisbury

Rwanda

Direction de l'Hydrologie, B. P. 323, Kigali
Service de la Météorologique, B. P. 720, Kigali

Senegal

Chef de Service Hydrologique, Sicap No. 2993, Dakar
Ministère du Development Rural et de l'Hydraulique, Dakar

Sierra Leone

Ministry of Works, New England, Freetown

Somalia

Civil Aviation Dept., P. O. Box 310, Mogadiscio
Ministry of Mineral and Water Resources, P. O. Box 639, Mogadiscio

South Africa

Department of Water Affairs, Private Bag 156, Pretoria
National Institute for Water Research, Ausspannplatz, Windhoek, S. W. Africa
University of Witwatersrand, Dept. of Civil Engineering, Johannesburg
Weather Bureau, Private Bag 97, Pretoria

Sudan

Geological Survey Dept., P. O. Box 410, Khartoum
Ministry of Irrigation, Wad Medani
National Council for Research, P. O. Box 2404, Khartoum
Sudan Meteorological Service, Gen. P. Office, Khartoum

Swaziland

PWD, Hydro, Box 58, Mbabane

Tanzania

East African Meteorological Department, P. O. Box 3056, Dar es Salaam
University of Dar es Salaam, Bureau of Resources Assessment and Land Use Planning, P. O. Box 35049, Dar es Salaam
Water Development and Irrigation, P. O. Box 35066, Dar es Salaam

Togo

Direction Génie Rural, B. P. 1468, Lomé
Service Hydrologique, B. P. 1463, Lomé
Service de la Météorologie, Lomé

United Arab Republic

Desert Institute, El Mataria, Cairo
Ministry of Irrigation, Cairo
Physical Meteorology Division, Egyptian Met. Authority, Cairo.

Uganda

Department of Water Development, P. O. Box 19, Entebbe
Meteorological Dept., P. O. Box 7025, Kampala

Upper Volta

Météorologique Nationale, B. P. 576, Ougadougou
Section Hydrologique, Service de l'Hydraulique et Equipment Rural, B. P. 3003,
   Ougadougou

Zaire

Direction de Service Météorologique, Kinshasa
Ministére de l'Agriculture, Kinshasa
Ministére des Travaux Public et des Télecommunications, Kinshasa

Zambie

Meteorological Department, P. O. Box 200, Lusaka
National Council for Scientific Research, Water Resources Unit., P. O. Box Ch 158,
   Lusaka
University of Zambia, School of Civil Engineering, P. O. Box 2379, Lusaka
Water Affairs Department, P. O. Box 530, Lusaka

INDEX